Also by Jane Alexander

Command Performance

The Bluefish Cookbook,
with Greta Jacobs

The Master Builder,
adapted from Henrik Ibsen
with Sam Engelstad

WILD THINGS, WILD PLACES

WILD THINGS, WILD PLACES

Adventurous Tales of Wildlife
and Conservation on Planet Earth

Jane Alexander

Alfred A. Knopf

New York | 2016

THIS IS A BORZOI BOOK
PUBLISHED BY ALFRED A. KNOPF

Copyright © 2016 by Jane Alexander

Grateful acknowledgment is made to Parallax Press for permission to reprint an excerpt from "Please Call Me by My True Names" from *Call Me by My True Names: The Collected Poems of Thich Nhat Hanh,* copyright © 1999 by Unified Buddhist Church Inc. Reprinted by permission of Parallax Press, Berkeley, California (www.parallax.org).

Library of Congress Cataloging-in-Publication Data
Names: Alexander, Jane, [date]
Title: Wild things, wild places : adventurous tales of wildlife and
    conservation on planet Earth / by Jane Alexander.
Description: New York : Alfred A. Knopf, 2016. | "A Borzoi book." |
    Includes bibliographical references and index.
Identifiers: LCCN 2016007803 (print) | LCCN 2016011120 (ebook) |
    ISBN 9780385354363 (hardcover : alk. paper) | ISBN 9780385354370
    (ebook)
Subjects: LCSH: Wildlife conservation.
Classification: LCC QL82 .A424 2016 (print) | LCC QL82 (ebook) |
    DDC 333.95/416—dc23
LC record available at http://lccn.loc.gov/2016007803

Front-of-jacket images: (top) Bison and (bottom) Rhinoceros, from *Mammals Illustrated from Nature* by Johann von Schreber. Both, De Agostini Picture Library/Getty Images; (background on animals) *The Great Valley* by Paul Weber (detail) © SuperStock/Alamy
Jacket design by Janet Hansen

Manufactured in the United States of America
First Edition

To Ed.
Thanks for the journey.

Contents

PART 3: THE BODY OF THE EARTH

Author's Note

My decision to capitalize the names of all animal species breaks with tradition and will cause grammarians to wince, if not throw these pages up in despair at what may seem capriciousness on my part. However, there is precedence for my action. The birding world has long capitalized the names of birds, making it clear that we are referring to the Solitary Sandpiper, the Great Egret, and the Black Phoebe and not descriptive terms such as "A solitary sandpiper waded along the shore," "We saw a great egret today," or "A black phoebe sat on a fence."Capitalizing minimizes confusion about the world's ten thousand species of birds, but it also elevates their status on a page, as capitals are intended to do by calling attention to the word. This is my intention in the book: to call attention to species of mammals, fish, reptiles, amphibians, and insects as well as birds. I have capitalized individual species but not families or categories. A single species such as the Tiger, with subspecies such as the Bengal Tiger, or a dominant species of several species such as the African Elephant or the Grey Wolf, will be designated by the generic "Tiger," "Elephant," or "Wolf" once it has been established in a chapter about Asia, Africa, or North America respectively. If the capitals are annoying I beg you to take a look at the sports page of any newspaper where teams, leagues, cities, coaches, awards, and players are all in capitals, adding considerable ink to any article. Surely the animals of the field and forest, the waters and the skies, deserve as much attention.

Prologue

An impenetrable wall of jungle vine cloaked the forest trail as we descended over a small stream and into even darker afternoon light ahead. The acrid scent of a large animal wafted toward us, and I stopped. If this was a Jaguar I didn't want to risk cornering it. Jaguars are not known to be aggressive toward humans, but I was taking no chances, so with a soft footfall we retreated and made our way back up to our cabin in the mountains.

Belize in 1982 was not the tourist mecca it is today. The country had just won independence from Britain a year earlier, and its abundant wildlife was known only to scientists, a handful of birders, and dedicated reef divers. A friend of mine in the Foreign Service was stationed there, and at her urging my husband, Ed, and I paid a visit.

The portal through which all pass is Belize City. We walked through the rough-and-tumble town, birthed in the 1600s for the lumber trade at the confluence of rivers and streams spilling into the Caribbean. Haulover Creek split the town in two and was a disgusting body of water into which everything was thrown, including carcasses from the outdoor market. Bloated fish, coconuts, shoes, dead dogs, and yesterday's newspapers floated by as if consecrating them to the river and the sea meant an end to the horror. But the detritus never stopped. Mosquitoes, rife with malaria, were active day and night while flies sank into luncheon plates quicker than a fork. A woman sitting in a concrete alcove begged me to buy her baby. We didn't linger here.

A bush pilot we met in a bar flew us to a Mayan ruin called Lamanai, largely unexcavated then. The little seaplane put down in the "river of crocodiles" and was secured to the wooden dock. Tripping over extensive vines, we began to explore the ancient site. No one was around. Only the incessant cicadas and tree frogs, the occasional whoop of an Oropendola above its dangling nest or the buzz of insects broke the silence of centuries. The great pyramid called the Temple of the Jaguar was unrecognizable since the jungle had reclaimed it as her own. We made our way down crumbling steps and entered a narrow dark corridor, ducking to avoid bats as they flitted into open air. The beam of our flashlight revealed a huge stone head looming in the dark, its blue and red paint as bold as the day it was applied a thousand years ago. This great Mayan civilization was gone. It disappeared as the Olmec had disappeared before it, both as extinct as the Dodo bird half a world away. By the time the Spanish arrived, the Maya had been replaced by the Aztecs and then the Aztecs by the Spanish. Survivors of these great civilizations were subsumed into a new culture, like Darwin's finches, adapting to the environment in which they found themselves.

Ed and I traveled deeper into the interior of Belize, to the high pine forests, staying in a government "rest house" courtesy of our Foreign Service friend. We awoke each morning to the roar of Howler Monkeys, the sound penetrating the forest like an approaching freight train. Once I glimpsed a big male on a dirt road one hundred yards ahead, walking on two feet like a dark furry human being. Sasquatch lives!

The sheer abundance of industry around us was unlike anything I'd encountered in the forests of the northeastern United States, from Leafcutter Ants chopping up leaves and carrying them in long soldierly lines to their farms, to myriad colorful birds zipping by in pursuit of food or nest material. Belize is a birder's paradise, but it is also home to thousands of species of insects, amphibians, reptiles, and mammals, including the king of the forest, the Jaguar.

I went home and began writing a screenplay about a female biologist who tracks Jaguars in Belize. I contrived a story that involved the modern-day Maya, drug trafficking, and our intrepid heroine. What a great adventure, I thought, and what a great part for me! In the course of research on the cats I called the Bronx Zoo and was told I might want to talk to their field biologist in Belize, who actually was tracking Jaguars there. He had just been grounded because of head injuries sustained in a crash involving a single-engine plane, which had nosedived into the jungle floor, and he might be amenable to conversation.

The next day I hopped on a plane and by evening was having a drink in the Fort George Hotel with the thirty-year-old Alan Rabinowitz. He was bruised and battered but intrigued enough by my presence to invite me back to his study site in the Cockscomb Basin. We spent five days together, tracking Jaguars against the orders of his doctor, dissecting scat, measuring pugmarks, and talking extensively about wildlife. His shack was deep in the rainforest next to a few Mayan families and surrounded by a cacophony of birdsong.

I never saw a Jaguar in Belize, but one night as I lay on my cot, fixated on five scorpions clinging to the inside screen near my head, I heard a deep guttural cough outside, which thrilled me to the core. This was it. This was the wild elusive cat of Central and South America, revered by the Olmec, Aztec, and Maya, stalked by trophy hunters and fashionistas for its singularly gorgeous coat. This was the animal so rarely seen or heard that it existed in the realm of myth and divinity. The shack wall was all that separated us as I listened to the Jaguar in the forest of the night. I was hooked.

I abandoned my screenplay and turned instead to encouraging Alan to write his story—reality is always more interesting than any fiction, and Alan's life in the jungle of Belize was exciting and different. As a stutterer who talked only to animals he had a lot to say.

My life has been immersed in the world of make-believe ever since I began acting. It is a magical world I love and that summons the most imaginative parts of me. But the natural world holds more mystery and beauty than could ever be contained in one life, or vista, or creature. As Hamlet says to his friend, "There are more things in heaven and earth, Horatio, than are dreamt of in your philosophy."

Alan and I became close friends and we traveled to many places together. With his wife, Salisa, and my husband, Ed, we trekked the Annapurna range in Nepal's Himalayas; we visited his study site in Thailand hoping to glimpse the rare Clouded Leopard; we spied Bengal Tigers perfectly camouflaged in the grasses of India; we rafted a great trout river in Idaho; we stayed with the Lacandon Indians, never subdued by the conquistadors, in the depths of the Chiapas rainforest; and, finally, we saw numerous Jaguars lazing on cool riverbanks in Brazil's searing Pantanal.

My thirst for wild encounters and my passion for birds have taken me to unique places in the world in the company of remarkable field biologists. One cannot spend time with these heroes of conservation and not understand the crisis we are facing today with wild things and wild places.

The changes that are occurring to our planet Earth, as a result of human incursion, are happening with such rapidity that by all estimates we barely have a generation, maybe twenty years, to slow carbon emissions into the atmosphere before the results are catastrophic. Many bird and animal species are threatened with extinction. The concurrent events of global warming and rising sea levels, which mean the end for birds, will force mass migrations of people to viable climates and away from coastal flooding. Birds really are the canaries in the coal mine, portending what's to come for all species. Where they struggle to survive, people also struggle.

We have not begun to explore the wealth that biodiversity offers human beings. Our interconnectedness with other species directly impacts our own well-being both physically and spiritually. We depend on the benefits from nature to sustain our bodies and the solace of wild places to soothe our souls, but somewhere along the way we lost respect for nature. We lost wonder. We no longer consider the unique and living creatures of the planet as sacred or special beings like ourselves.

The killing of animals for body parts, or out of irrational fear or irresponsible hunting, and the collecting of rare species for private gain have become rampant and threaten the populations of the most magnificent mammals on earth, such as Elephants and Tigers as well as the less celebrated amphibians, reptiles, and fish.

We are desecrating our lands, ripping out the heart of precious sites through reprehensible drilling, logging, and monoculture. We are polluting our waters with chemical spills, plastics, sewage, and dredging. Nature is resilient; she can repair herself if given the chance and the time.

Climate change is the overriding issue of our generation, dwarfing all others. One hundred and ninety-five nations finally came together in December 2015 to commit to reducing carbon emissions, officially recognizing that we are one world and we sink or swim together. The hard work of actually making it happen has just begun. Can human ingenuity pull the rabbit out of the hat with this one? I have great faith in technology's saving us or at least leading the way, but I have less faith in our leaders in government, the media, and education making the case for our citizens, or teaching our young the importance and wonder of the natural world. How do you learn to love or protect the marvels of this planet if you have not been taught about them? How do you stand up for those things that have no voice of their own if you are not shown the way?

Scientists know what is happening, but their science is often repudiated and they cannot make their voices heard. They are like prophets in the wilderness. They have written thousands of books and papers. I have read only a fraction, but I have traveled with some of them and I have listened as they recounted the wonders and horrors of their journeys; I've also listened to the wildlife photographers who are documenting this precipitous time in our planet's history through their artistry. These men and women are my heroes, as are people everywhere who speak up for the natural world. This book is for the protectors of our home the Earth and for all the miraculous living things in it.

PART 1

TIGER MAN

1

Mexico

Shade trees spread their branches over walkways of the little zoo in Tuxtla Gutiérrez, a respite from the late afternoon heat. Alan Rabinowitz was ahead of us, looking down into a spacious enclosure that mimicked the real rainforest of Mexico to our south. We heard the guttural snarls before we saw the object of his intense gaze. A female Jaguar snapped at the male as he attempted to mount her, and swiped her paw across his cheek.

He grabbed her by the neck, thrusting her low to the ground, but she managed to escape and ran off into the underbrush with the male in pursuit. He sauntered next to her and began licking her back and neck against the lay of the fur, surprisingly gentle compared to the bite to her neck. She didn't stir until he again tried to get on top; then the whole courtship ritual began again. When at last she was ready he gained entrance and held her down with his jaws on her neck until the act was finished minutes later. They rolled on the ground and spooned like lovers on the grass, satiated for the moment. Then the process began all over again.

Alan was transfixed. He watched for the better part of an hour, barely moving, just like a cat. That is what Victor had said earlier in the day, when he first met Alan: "He *is* the cat he stalks!" It was true. His stillness was sometimes unnerving. His eyes, pale and catlike, could fix you with a penetrating stare while he contemplated the

Jaguar in a small zoo, Chiapas, Mexico, 1985

answer to something said. He was also powerful. He had excelled at wrestling in school and kept his body strong with weights and running. Later he would be dubbed the "Indiana Jones" of conservation for his intrepid pilgrimages on behalf of the great cats.

It was 1985. Alan, Ed, and I were traveling with Victor Perera in southern Chiapas. Victor, a Guatemalan, taught writing at the University of California, Santa Cruz, and had spent years with the Maya and the Lacandon Indians, a tribe never crushed by the Spanish. His book *The Last Lords of Palenque* painted a picture of the waning days of a once-great society, one that had built the graceful temple Palenque and worshipped gods and demigods of the animal world, including the Jaguar. I had studied the Maya, and Alan wanted to know more about the spirit world and the Jaguar. Victor's book was fascinating; I sent a letter to his publisher telling him so. Our correspondence resulted in a friendship and an invitation to accompany him on his next visit to the Lacandon, in the last great forest of southern Mexico.

Chan K'in Viejo, his body bent, one shoulder higher than the other, his hands gnarled from the arthritis afflicting him, stood on the hill near his thatched house as we climbed to meet him. He smiled warmly as Victor approached, and they embraced. He said he had dreamed we were coming. This last lord of the Lacandon was barely five feet tall; he wore a white cotton tunic, and his black hair fell straight to his shoulders, with bangs covering his forehead. His voice was strong, and he was full of goodwill and humor. Victor presented him with a box of Churchill cigars, and Chan K'in immediately lit up, the huge cigar covering half his wizened face as he smoked. We spent the afternoon and evening under his roof as his three wives ground flour in the metate and pressed tortillas onto the hot stones in the fire, his little children treading the dirt floor with their stick toys. Chan K'in was reputed to have twenty-one children and in his nineties still copulated on a ritual pile of corncobs after the annual harvest.

He showed us the God House, where the men participated in balché ceremonies. Balché is an ancient intoxicating drink made from fermentated honey and the roots and bark of the sacred Balché tree. The gods are invoked with chants and offerings and the inebriated men soon begin to receive messages from the spirit world through

Victor Perera, the Lacandon shaman Chan K'in, and me,
Chiapas, Mexico, 1985

hallucinatory visions. We stood outside looking up at the tall trees, one a giant mahogany used for hollowing out dugout canoes for the waterways. Looking at the colorful birds flitting by, I waved my arms in the air and said in my primitive Spanish, "Usted, mucho perros." Chan K'in politely nodded while Victor broke up. "You said 'You have a lot of flying dogs'!" We had a good laugh.

There was something magical about being in Chan K'in's presence; he radiated wisdom, and we felt good. Victor told us that strange things happened here, that any technological device would break down after awhile. Alan and I were skeptical and I turned on my tape recorder at the conclusion of a dinner of rice, black beans, and the freshly made tortillas. Alan asked about the Jaguar, the "Were-Jaguar" and the "Master Jaguar," while Victor translated, sometimes in Lacandon and other times in Spanish. Chan K'in began to talk.

He said they believed in a Master Jaguar that leads all the other Jaguars and in a minor god who transforms himself into a Jaguar to roam the forest. It was not clear if they were one and the same. "How do you know it's the Master Jaguar?" asked Alan. "Because he speaks to you," said Chan K'in. This resonated with Alan. In Belize he had sometimes heard his name being called in the night as if spirits were summoning him. He had also had an experience with an Obeah man who practiced a kind of black magic and who told him after a séance that a Jaguar would appear in his trap within three days. And there, on the third day, was a Jaguar.

The Mayan belief in the spirit world was clearly in decline as more Maya were converted to Christianity; still, a residual belief and even fear of a guardian of the forest persisted for many. As long as people believed in a controlling spirit or demigod, more protection was offered the animals of the forest. There was fear of retribution if one hunted too many animals or did not pay attention to the signs the spirit world gave, such as being bitten by a snake. Yet the arrogance of the Judeo-Christian ethic, placing man at the top of the pyramid of life rather than on a branch of the sacred Mayan tree, was dislodging the ancient beliefs of these people.

Alan asked Chan K'in if the voice in the night that called his name was the Master Jaguar. Chan K'in looked at Alan, his eyes shining in the firelight, and Victor whispered, "That's your secret, Alan." My tape recorder broke after a half hour. And then the light meter in my camera stopped working, as did my little travel clock. I have no explanation for this except to acknowledge how much I do not know of this world.

We bedded down in a tiny shack, Victor and Ed in hammocks and me pretzeled on a four-foot-long shelf with tiny spiders scurrying for safety. Alan spent time alone in the God House and then curled up in our VW van. He had endured enough nights battling insects, bacteria, and the chatter of sleepless jungle families.

In the morning we bid goodbye to Chan K'in. He gave me a very special gift of a god-pot, in which copal was burned for the Balché ceremonies. He was a most remarkable man to have graced our lives. Twenty miles down the road we got the tires of our van pumped up at a huge sawmill that was the repository for all the logging of the Lacandon National Forest. There, piled as high as a three-story building, were the giant mahoganies and the ceibas, the sacred tree. As far as the eye could see, the land had been cleared for cattle ranches and agriculture. Most of the Lacandon men had signed a government contract allowing extensive logging of the forest in exchange for money. Chan K'in was a holdout, explaining that the forest was not his to give away. "I didn't plant the trees," he said. "They're God's," meaning they belonged to the great Lord Hachakyum, the major Lacandon deity. "Go ask him." The greatest rainforest in Mexico, supporting a third of all the biodiversity in the country, was straining to stay alive. The fragmented landscape kept chopping away at the wildlife.

We walked the ruins of Palenque, marveling at the artistic line of stone Jaguars, safe for another thousand years for tourists like ourselves. Evening found us at Na Bolom, the home of Trudi Blom, the widow of archaeologist Frans Blom, who devoted her life to helping the Lacandon preserve the forest and their way of life. Trudi reputedly said,

I have learned through bitter experience that you cannot hope to protect the Lacandones without safeguarding their forest . . . In the dreams of the Lacandones, which regulate their waking lives, each animal, each plant and each ritual object is an instrument of prophecy or protective magic. As the forest is burned and cut down through our stupidity and greed, the animals disappear one by one; the jaguar, the boar, the puma, the spider monkey—they all disappear, and soon the souls of the Lacandones will also disappear . . . it makes no difference how many of them will be left—the fact is, their souls will wither and die as their magnificent forest is destroyed, and all of us will share part of the blame . . . What you are seeing is the last whisper of a magnificent culture . . . But in Naha, if you gain the confidence of Chan K'in Viejo and other elders, you can still get an idea of what the Lacandon culture was about. Chan K'in in particular is an extraordinary man. So far, he has not permitted the government to cut mahogany around Naha. When he dies, there will be no stopping them.

Chan K'in died on December 23, 1996, at 104 years old. The Lacandon forest continues to be logged. No one knows how many Jaguars still exist there.

2

Belize

Nothing in Alan Rabinowitz's early years would lead you to believe that he would become one of the great conservation heroes of our time, except for the solace he always found in the company of animals. He was born on New Year's Eve 1953 in Brooklyn, during the worst snowstorm of the century, it was said. He grew up a quiet child in Far Rockaway, placed in a special-education class because of a severe stutter. Alan found comfort sitting in a dark closet with his pet turtle or chameleons because there he could talk to them without stuttering. His father, a vigorous high school coach known as Red, noticed his son's affinity for animals and took him to the Bronx Zoo on free afternoons. Alan gravitated to the Lion House, where the big Jaguar paced back and forth in her lightless cage, on a path worn into the stone by her short miserable journey from one wall to the other. The little boy stood mesmerized on the other side of the bars; here was a trapped, voiceless creature like himself. He promised her that if he ever found his voice he would be a voice for her and keep all animals from harm.

Years later an auspicious meeting with George Schaller of the Wildlife Conservation Society (WCS, then called the New York Zoological Society, the umbrella organization of the city's zoos) changed his life forever. Schaller was working on Giant Pandas at the time and Alan was finishing up his PhD at the University of Tennessee when a visit by the great field biologist resulted in Alan's taking him on a hike to compare the Black Bear habitat of the Smoky Mountains with that of Pandas in China. Schaller was the head of the International Program of WCS and needed someone to study Jaguars in Belize. He recruited Alan, and soon Alan was on his way to Central America and a life-changing experience.

Dr. Rabinowitz was well into his two-year study when I flew to Belize to research the movie proposal I'd written about a female zoologist. He suffered headaches from the recent plane crash. The single-engine aircraft had plummeted from the sky while he was tracking his radio-collared cats from the air. The propeller ground into the tangled forest floor, but Alan and the pilot managed to extricate themselves and walk away from what seemed like certain death. His head injuries needed time to heal, so it was a good time for me to visit.

The young field biologist was recovering from the crash, but he had been weary for some time of the immense efforts it took to do his work. The magic he had felt when he first came into the forest, the cacophony of insect life, of birdsong, the mammals and reptiles that he encountered on forest treks, the glory of plenty, had faded in the wake of constant trials. The difficulty of capturing and then safeguarding the Jaguars, the relationships with the Maya and the women in his life, the constant assault on his health from parasites and injuries—all these were weighing on him.

My arrival was serendipitous. I was immediately entranced with the rainforest. Most of my prior forays to the tropics were typical Caribbean vacations, lazing on beaches, a mai tai in one hand and suntan oil in the other, some hikes here and there and visits to the requisite tourist sites.

When Ed and I went to Belize in 1981 to visit my Foreign Service friend, Cynthia Thomas, I realized just how vast a world there was off the beaten path. Our world of theater and film, of imagination and the magic of illusion, never ceased to excite us. We would jump into a new project and know we were the luckiest people on earth to be involved in the exploration of human behavior, emotion, and the mind. There were layers upon layers to be exposed and the process never ended. But venturing into the wild, into landscapes dominated by animals, onto paths trod by few other human beings, presented even more exciting worlds of exploration.

At Alan's study site, deep in the jungle of Cockscomb Basin, I was in my element. An insect species the size of a hummingbird must have just hatched because pale orange bodies filled the air by the hundreds and emitted a whirling hum, like toy helicopters. No one could tell me what they were, and I have never seen them anywhere since. A Laughing Falcon perched above the porch, cackling at my intrusion, antbirds followed the legions of army ants on the move, and dozens of different kinds of flycatchers whistled in the forest. The sheer abundance of life was mind-boggling.

Tropical forests cover about 7 percent of the earth and hold more than half of the animal and plant species. No one knows how many species exist on earth, but the figure could be as high as thirty million. They are still to be discovered. These forests used to cover as

Cockscomb Basin, 1984

much as 20 percent of the earth's surface, but as human beings continue to convert the land to agricultural and pastoral uses more species are lost every day.

The tropical rainforests are a delicate balance of soil and climate, where the temperature range is usually between 64 and 100 degrees Fahrenheit and the precipitation is not less than 66 inches annually. Whatever nutrients are in the soil are mostly absorbed by the biomass of trees and plants, leaving the soil itself rather barren when the forest is demolished. Speciation flourishes in rainforests for a complexity of reasons not fully understood. But the farther one gets away from the equator, the fewer species exist. In Belize almost six hundred species of birds can be seen. By the time one reaches Antarctica there are about forty-five species, mostly pelagics, which spend most of their lives at sea.

In those first five days with Alan in Cockscomb when we were getting to know each other, we walked the trails looking for Jaguar scat or pugmarks. Once he told me not to stop but to calmly keep walking on as he sensed a Jaguar a few feet off the path. He knew the forest and pointed out the secrets of medicinal vines and one that streamed water from its core when sliced, just in case one needed a thirst quencher on the trail. He was a restless patient and would not

Alan Rabinowitz listening for Jaguars he has radio-collared

stay in his shack. At night I accompanied him when he went to track the cats on a starlit rise, earphones on, the antenna held high in the air to catch the beeping radio signal. When there was none he worried for his Jaguars as if they were his children.

We read poetry to each other on rainy evenings, and one afternoon he offered me his journals to read. These were not simply data listed at days' end by a scientist. These were the musings of a young man documenting his rite of passage from naïf to sophisticate, from pure scientist to committed conservationist. There was a writer in these highly personal pages. He managed to balance fact, storytelling, and emotion, something I hoped for in my work as an actress. I told him he had a book in the making and that his was a hero's tale, the struggle for the Jaguar's survival. He demurred, and, changing the subject, read me his favorite poem, Robert Frost's "The Road Not Taken":

> *I shall be telling this with a sigh*
> *Somewhere ages and ages hence,*
> *Two roads diverged in a wood, and I—*
> *I took the one less traveled by,*
> *And that has made all the difference.*

What was the other road? I wondered. It would still have encompassed exploration of some sort, he said. Not the arts but medicine, perhaps, as he was a premed student in college.

In New York many months later, I introduced Alan to a literary agent, and his first book, *Jaguar*, published in 1986, became a classic of nature writing and a young man's inner journey.

His study site in a former logging camp was near several Mayan families. The women beat clothes on washing rocks in the river and I would join them with their children in the water. The little girls attached themselves to me like puppies, giggling when I tried out my Mopan or Spanish. One girl in particular stood out. Seven-year-old Agapita carried a child's book around with her but only half of each page, the other half gone forever. Still, it was her treasure and

The last of the unconverted

she was learning to read the words. She was curious about everything. There is always one child who breaks the mold. I've observed the same about animals: one will be more daring, or clever. Perhaps this is how speciation begins, through subjective behavioral change.

The water was beautifully cool on these mornings when the heat seemed to exude from every leaf and vine around us. Most of the women wore the traditional huipil blouses, but one, the youngest mother in her late teens and the newest arrival in the community, was naked above the waist. This was a village in final transition. Missionaries had taught the women that it was not good to be naked, so they clothed themselves. Jesus was a welcome god because sins were forgiven and he gave them hope for an afterlife. What was lost was a connection to all the gods and demigods of animism, a direct line to the creatures of the forest. Would Jesus protect *them* too? Yes, God the Father almighty, maker of heaven and earth, would take care of them all. The responsibility in this mortal life was relinquished.

Alan did not set out to be a conservationist. He was there to research the top carnivore of Central and South America, a cat that had roamed the New World for perhaps a million years. The Olmec of three thousand years ago, the Maya and the Aztecs later, all revered the Jaguar, as represented in their art and writing. There are stunning Olmec statues of men holding limp Were-Jaguar children, the cross between a Jaguar and a human, which some speculate was indicative of child sacrifice. Thousands of Mayan ceremonial vessels depict Jaguars, and there are references to them on stone stelae and in the Popol Vuh codex. The Aztec had an elite military unit called the Jaguar Warriors, who dressed in Jaguar skins to incorporate the strength of these gods of the night.

There were perhaps ten million people living in Central America before the Spanish conquest. Who knows how many Jaguars there were and how they managed to survive the people who hunted them even as they revered them? But they did survive and continued to disperse successfully throughout South America. Even before the Spanish came, the prolific Mayan cities collapsed, by 900 AD, probably from drought, warfare, and disease, the great temples left to decay in the sun or be swallowed by the jungle. The Spanish conquistadors who arrived in the sixteenth century brought smallpox, measles, and the flu, effectively wiping out most of the existing population. The Jaguar became king of the forest again. But not for long—five hundred years is not long in the scheme of things. Modern guns entered the picture, and by 1980 Jaguars were gone or severely diminished from Arizona south throughout Mexico.

Alan trapped seven Jaguars in his years in Belize in the 1980s. He tranquilized and radio-collared them, measured them, and medicated their infections and parasites before they awoke and disappeared into the dark forest again. He learned a lot: the density of the animals in the area, what they ate, and how far they traveled. He pioneered research on the species. But none of it was easy. Roads washed out regularly after torrential rains that sometimes lasted for weeks; his truck kept breaking down; and the well became contaminated when no one noticed a dog's rotten carcass in it.

The profusion of animals in the tropics includes, of course, the tiniest ones: bacteria, amoebas, and parasites like hookworm. Mosquitoes are ever present and a bite can result in malaria or yellow fever, while a deposit of botfly larvae burrows under the skin and begins to eat you inside, causing stabbing pain before the fully grown two-inch worm finally exits months later. You learn not to grab onto trees or vines for fear of thorns or biting ants or an insect that mimics the bark or a snake the green vine.

Underfoot the forest has many wonders and as many dangers. The deadly Fer-de-Lance snake—or "Tommy Goff," as it is known locally—lies in its hole near the base of a tree and will defensively strike as you pass by its territory. You learn to walk gingerly around protruding roots for fear of a snake hole. Alan lost a man to a Fer-de-Lance bite when they were out trapping his first Jaguar. Guermo was wearing sneakers, and the snake bit him in the ankle. Victims of a Fer-de-Lance bite experience excruciating pain, internal bleeding, and swelling; if there is no treatment, death can occur within hours from hemorrhagic shock. Guermo received antivenin and antibiotics in the hospital where Alan urged him to stay, under the care of Western medicine. His family took him home, however, where he died a few days later.

Alan became close to the Mayan families of Cockscomb, ministering to their medical needs and welcoming them into his shack at all hours. He was known far and wide as "El Hombre Tigre"—Tiger Man. (The Jaguar was known as El Tigre.) There were many stalemates involving the people and their relationship to the wildlife of Belize. It was illegal to kill the big cats and yet they did it anyway,

Alan writing in his journal, Cockscomb, Belize, 1984

because the animal was too near cattle, or because they could get $100 for its skin, or simply because "it was pretty" and they wanted to nail the dead Jaguar to the wall rather than see its black rosettes glowing on a riverbank in the late afternoon sun. Alan's frustrations coalesced into depression, then anger, and, finally, a single-minded motivation to change the equation. He would find a way to protect the Jaguars in the Cockscomb Basin by making it a preserve. The voiceless young man began to find his voice.

Conservation was not something known to the Belizean government in the 1980s. They won independence from England only in 1981 and Belizean politicians were still finding their way. The Belizean Audubon Society was formed in 1969, however, as an offshoot of Florida Audubon. The abundant bird life in Belize was well known to birders, and Belize Audubon was interested in preserving important areas in the country. Alan had his first partners in BAS for the preservation of Cockscomb.

If saving Jaguars in the forest was frustrating, saving them on paper was equally so. Political life is a waiting game, a game of compromise and broken promises, and the learning curve is steep. Alan kept at it despite roadblocks. The folks at Belize Audubon one day warned him that some of the Maya of Cockscomb had applied for land titles. They had been squatters for years in the forest just as their ancestors had been, the concept of ownership being as alien to them as it was to Native Americans to the north. Missionaries told the Maya how to go about getting the land so they could build a new church in their village, a move clearly incompatible with a Jaguar preserve.

Alan knew these missionaries. He had a run-in with them when he saw their children killing birds with a slingshot. He told them the area might soon be a preserve and asked them to tell their children to stop killing the birds. Their only reply was "We save souls." Later they came into Alan's study site and showed the Maya a movie of sinners burning in hell, telling the Indians they had to repent. They also told them they could have a nice new church if they got legal title to the land. Alan ramped up his visits to the capital.

There are always winners and losers when lands are locked up for private enterprise, commercial development, military uses, or wildlife refuges. Alan sympathized with his Indian friends, but Jaguars were more plentiful in the Cockscomb Basin than anywhere else, and if they were protected, then the entire forest would be protected in perpetuity. It was time to put the animals first.

His position put him in danger from other competing interests. Hunters expected to continue their illegal killing and threatened Alan with bodily harm. Marijuana growers likewise expected to continue to harvest their crop. He had no police to call on in the jungle, just his unwavering resolve to protect the animals and their land. It worked. It took more than a year and many meetings with forestry and natural resources representatives in the government but finally, with the constant help of the Belize Audubon Society, the Cockscomb Forest Reserve/Jaguar Preserve came into being in December 1984. One of Alan's closest Mayan friends, little Agapita's father, Ignacio Pop, became its first warden. Alan left the Belizean forest a short while later. He had managed to protect the Jaguar in this small part of the world where Belize is still 70 percent forested and it is possible to see the cats in the wild. But the idealism he brought to his research as a thirty-year-old biologist was gone. It was not enough to understand the animals, even to love them; it was incumbent on him to protect them, the very promise he had made to the Jaguar in the zoo as a little boy. The future for wildlife was about war, and he knew the fight would be hard.

3

Thailand

The girls had no expression on their beautiful faces as they slid suggestively up and down the poles in the oscillating blue and amber light. They flicked their long black hair round and round and pushed their tiny breasts and buttocks forward and back. Beefy Germans lifted large steins of beer and drunkenly catcalled to them above the canned music. One of the men pointed at a girl, and she teetered in her high heels down the platform's side steps and joined him at a door in the back of the room.

Patpong is the famous red-light district of Bangkok where parents in rural villages often send their teenage daughters to bring back money for the family. I was struck by the tender age and indelible beauty of these Thai girls. Few women in the world are as beautiful as Thai women. They are delicate and perfectly proportioned, with copper skin and thick black hair. It was appalling to see them being targeted for sex.

How in the world would it be possible to protect wild animals if human beings were treated with such degradation?

This was our discussion as Alan, Ed, and I traveled from the choking traffic of Bangkok to the Huai Kha Khaeng Wildlife Sanctuary, where Alan studied the Clouded Leopard and other cats near the Burmese border. A week earlier Ed and I had visited a Cambodian refugee camp with our friend Cynthia Thomas of the Foreign Service. By coincidence she and Alan were again working in the same country, first in Belize and now in Thailand in 1990.

The killing fields of Cambodia under Pol Pot were horrific. No one knows how many died in the genocide between 1969 and 1979, but some say as many as three million, out of a total population of seven or eight million. Border camps were set up all along the eastern half of Thailand, presenting a huge challenge to the international humanitarian community. They existed for twenty years, like Cambodian cities. Hundreds of thousands of refugees lived far better inside the camps than they would have outside.

Evelyn Sak, the wife of General Sak Sutsakhan, the last head of state of the Khmer Republic before it fell to the Khmer Rouge in 1975, was determined that the generation lost to the killings and camps would somehow not lose its ancient and highly evolved art forms, particularly Cambodian dance. When she learned from Cynthia that Ed and I were theater and film professionals, she arranged for the remaining members of the classical dance company of Phnom Penh to perform for us.

We sat facing an open-air stage on a warm February morning in 1990. Little children whose legs had been blown off by land mines that riddled the Cambodian countryside sat on the ground in front of us with their crutches and makeshift prostheses. Disfigurement from the mines was commonplace in the camps. It took decades before the land mines were removed from the fields and traveling was safe again.

A handful of musicians began playing gong chimes, drums, and bamboo flutes as the female dancers took the stage. They had salvaged some of their silk-brocade bodices, and with black kohl around their eyes and elaborate headdresses of gold they danced as if, like their ancestors, they were performing for the royal family. Bells shook at their ankles as they lifted their knees; their toes and fingers seemed to extend to impossible lengths, curving skyward like half moons. The children were rapt. The horror visited on these innocents juxtaposed against the glory of this human endeavor was a lesson in hope. Human beings are amazingly resilient.

Evelyn Sak and I kept in touch, and a few years later in 1995 when I was chairman of the National Endowment for the Arts under President Bill Clinton we met again in Long Beach, California, where many of the refugees settled after release from the camps. The NEA awarded a grant to the Cambodian dancers, and the ancient art form continues today, passed down to a new generation of dancers in Cambodia's capital city of Phnom Penh by these very survivors of the refugee camps.

The degradation endured by people the world over at the hands of other human beings was very much on our minds as we drove north with Alan to his field site, fittingly called Dancing Woman Mountain. Cambodia was to the east and Burma to the west, countries that had suffered ethnic and political conflict for many years. The Burmese military placed the popular Aung San Suu Kyi under house arrest in July 1989, just six months before we journeyed to Thailand, putting the political turmoil there very much in the spotlight.

Thailand is also a politically unstable country, despite being a constitutional monarchy with a prime minister. It has had twenty charters and constitutions since 1932. Theravada Buddhism deeply marks the character of the country and its people. But while all living things are respected, the ancient culture of killing still decimates wildlife. Thailand, like most Asian countries, has traded in wild-animal body parts for hundreds, perhaps thousands of years. Now the commercialization of that trade and the continual destruction of the forests threaten most of the great mammals, birds, and reptiles with extinction. On paper Thailand has prohibited trade in animal parts; in reality, Thailand is the second largest portal of that deadly trade after China.

Alan was asked by Thailand's Royal Forest Department to do a study of the animals of the Huai Kha Khaeng Wildlife Sanctuary in hopes of saving this last best place from illegal hunting, tribal people, and the rich and powerful Chinese. By concentrating on the density of wild cats there he could learn the density of prey species as well; by protecting the top "landscape species," the habitats of other species would be protected by default.

By the time we arrived, his fieldwork was over and he was writing his assessment for the government. It had been a difficult few years. The initial intrigue of the people and the place, the early enthusiasm with which he had greeted his assignment, had eroded, and we saw a man in some despair. The few cats he captured and collared all died. The first, a Leopard, died from injuries sustained during capture, and he never saw his main study animal, the Clouded Leopard, alive, only dead as victims of illegal hunting. The trade in body parts for the medicinal markets of China, Thailand, and Vietnam in particular was ongoing. The authorities paid lip service to protection, but corruption was ingrained and widespread.

The Thai are complicated people. They claim respect for all sentient beings and yet they seem insensitive to the pain of animals. Many times Alan witnessed them skinning snakes or frogs alive or roasting live fish on the fire, when it would have been simple to end their suffering with a quick slit to the spine. When he suggested that one woman kill the frogs first before cutting them up, she replied, "It is not right for Buddhists to kill."

One Buddhist precept urges giving to others as a way to imbue the giver with a sense of generosity toward all things. This includes "merit making." One way to earn merit is to release captive creatures. This is laudable when it is shrimp or fish stuck in mud after a drought and one helps them back into the water. But the Thai markets are crammed with caged birds, turtles, and reptiles, caught or raised only to be bought and then released to confer merit on the giver. Often they are recaptured by the vendor and sold again. This kind of convoluted thinking extends to many practices in Thai life, and can be difficult logic for a Westerner to understand.

As in markets all over Asia, Africa, and South America, anything can be had in the markets in Bangkok: birds, monkeys, lizards, pythons, legal and illegal. The illegal ones are just in the back rooms or under a table and it only takes a little coaxing or a bill in the palm to see what you want to see—or don't want to see, as the sight is pitiful. The animals are commercial products treated with about as much care as a wooden spoon or a metal pot. If they die in the hot sun, so be it. That's the luck of the draw.

I once witnessed an event on a crowded Bangkok street that for me epitomized the paradoxical nature of Thai people. We were stuck in traffic in a tuk-tuk, one of the pedicabs so prevalent as a mode of transportation. At last the traffic surged ahead when suddenly a motorcyclist in front of us was slammed by a taxicab, smashing him and his bike to the pavement. It looked bad. Everything stopped around us as the driver leapt out of the cab and came round to the still rider on the ground. He gently touched him, whereupon the cyclist slowly staggered to his feet and, laughing, grabbed the cabdriver in a tight embrace. They jumped up and down together and danced in the street, exuberant that he had escaped death. The luck of the draw.

Jubilation is not common to see; Thais hold their emotions close to their chest, as Buddha teaches. The social structure would be disrupted if anyone exhibited extreme emotion, even affection. The repression of that emotion, however, often leads to outbursts of violence and dependence on drugs and alcohol. Alan is a quiet respectful man who always seeks to know the culture in which he finds himself, but when he is consistently let down his anger erupts. The laissez-faire attitude of the Thais clashed with his Western impatience to get things done. On Dancing Woman Mountain a bridge was not repaired after Alan's repeated requests and his truck fell through, injuring his hand. Alan was openly furious, a display of anger for which the Thai men never forgave him.

But it was the ingrained culture of killing, the almost schizophrenic attitude of Buddhist ethics juxtaposed with a cavalier insensitivity to wildlife, that was hardest to comprehend. Thailand passed the Wild Animal Preservation and Protection Act in 1960, putting nine species off-limits to all hunting and custody, including the Javan and Sumatran Rhinos, and native goral and serow, a kind of goat/antelope. The Rhinos were soon extinct in the wild, and the serow and goral barely hung on—despite the legislation, the hunting never stopped. In 1983, Thailand signed on to the Convention on International Trade in Endangered Species (CITES) but continues to launder exotic and endangered species through its borders and into its shops, rivaling China in this lucrative illegal activity. By 2014, Thailand was importing Elephants from Burma to satisfy the tourist industry's need for entertainment, and slaughtering Tigers and other wild cats to supply the Asian trade in body parts. During his time there, Alan began to comprehend that the poaching of wildlife in the forests was simply the underbelly of a vast network of corruption that went all the way to top government officials.

Dancing Woman Mountain was on fire, literally, when we arrived late in the afternoon at Alan's cabin. It was not the roaring kind of fire you see on the news, where California's mountain homes are being burned to the ground. This was a slow ground-level fire, which licked the trunks of the dipterocarp trees, a species tolerant of flame.

These fires were historically natural occurrences but are now man-made in many parts of Asia. They are set to promote new plant growth, expand communities, and keep areas clear. We wondered what happened to all the ground species in the fire's path, and presumed that snakes, frogs, turtles, and other animals with nowhere to go simply perished. It was unsettling to see the fires just thirty feet from the house as we retired for the night, but we were told that there were fire breaks to keep it at bay.

Ed and I were given Alan's double mattress on the floor of his room for our stay. I shared my corner of it with an eighteen-inch skink who slept behind a laminated map nailed to the wall just inches from my body. Gravity slid him down to the bottom of the map, his belly in line with mine or my back. It took a little getting used to, but he barely moved all night, and when I did get a glimpse of him one afternoon as he was on insect patrol, I saw he was quite handsome in his mantle of turquoise and green.

Dancing Woman Mountain was beautiful, filled with tall trees and stands of bamboo. The air was fresh after the fires burned out and a gentle breeze blew through the camp. In the early morning saffron-robed monks came out of the forest and stood in a line on the

Accepting the offering of food,
Dancing Woman Mountain, Thailand, 1990

path near the workers' huts to have their bowls filled with rice, the act of giving conferring merit on the giver. Then the monks disappeared again to meditate or, if they were lucky, to come face-to-face with a Tiger deep in the forest, which was tantamount to meeting the Buddha himself.

I rose at dawn with the birds one day and walked down the bumpy dirt road to an incline, where the sun was peeking through the trees. I knew I would not see a Tiger, as Alan had found only one resident cat in the entire sanctuary. Leopards were more common, but Alan saw them only in his traps or as skins villagers hoped to sell him. It was the birds I was interested in as I sat on a log and watched the sunlight penetrate the wispy fronds of bamboo. Swifts circled the sky above me and a Sultan Tit, dressed in black with a lemon-yellow crest, hopped among the branches, scolding in its reedy voice, while three kinds of orioles fluted from the canopy.

A troop of fine-boned Rhesus Monkeys chattered to each other as the smallest ones climbed to the very tips of the bamboo to munch on new buds, the stems arching over the ravine with their weight. They were backlit against the horizon like a Japanese painting. I kept still, and they never knew I was there. This was a bit of paradise.

Ed and I trekked up the steep mountainside following a young man who promised the sight of a Rufous-necked Hornbill. Here in Thailand they were almost gone, hunted for their feathers and the great bill. We were careful not to slip on the occasional pile of steaming dung, courtesy of a rare herd of Asian Forest Elephants ahead of us, and three hours later we arrived at a raging boulder-strewn river. Our guide pointed and there, sixty feet above in the largest branch of a tall straight tree, was the hornbill on her nest. She lifted her heavy body and extended her wings a full five feet, soaring over us, with her curved black and white bill leading like the prow of a ship. Indelible sight.

Alan was accompanied by a smart and beautiful geneticist named Salisa, with whom he had been spending time in Bangkok. We ate together in a spacious hut where meals were prepared. Thai food is always delicious whether it is sold on the street or in a fancy restaurant. One evening the new chief of the sanctuary, Seub Nakasakthien, joined us. Seub was a senior wildlife officer in the forestry department. His dedication to wildlife was well known, and if anyone could put an end to the rampant poaching, it was Seub. The

Rangers heavily armed against poachers,
Huai Kha Khaeng Wildlife Sanctuary, Thailand, 1990

level of violence had escalated during Alan's time; there was now an all-out war between the poachers and the forest guards, resulting in the death of two of the guards. Everyone in camp carried a weapon for their own safety, while animals continued to be killed at an alarming rate. The workers estimated that as many as sixty poachers were in the sanctuary regularly. They were tribal people, mostly Hmong, who killed for food and anything else that could bring in money, but they were also police and soldiers who killed for the trade in body parts. Most of the catch was clearly illegal, and they knew it.

Seub was a decent, honest, intelligent man with a passion for the wildlife of his country. In his first six months on the job he arrested more poachers than had ever been arrested before, working in the field alongside his forest guards. He introduced many new research projects in the sanctuary and proposed it as a World Heritage Site. But the same people in forestry who had urged Seub to take the job were the ones playing a double game, the ones who gave him no support for his efforts. The poachers threatened to kill him, and he began wearing a bulletproof vest. His anger increased and depression spilled over into despair. On September 1, 1990, Seub Nakasakthien put a bullet through his head.

There are many eco-martyrs like Seub in the world today. They put themselves on the front lines for the preservation of wild things and wild places and they pay with their lives. They get jailed or killed, or, like Seub, they take their own lives in despair at the intransigence of corruption that defeats them.

Despair is easy to feel in the wildlife wars. Dead animals—animals that have been eviscerated for their tusks or gallbladders or bones or horns or fur—lie immobile, witness to our communal failure as human beings. If all people had enough to eat, if we consumers were not so greedy, if we celebrated the glorious variety of all living things and understood how interconnected we all were . . . if, if, if.

The contrast that is Thailand, that allows the snake to be skinned alive and the monk to look into the Tiger's eyes, is the deep heart of Buddhism and what it means to be human.

The Vietnamese Zen monk Thich Nhat Hanh describes the

dichotomy eloquently in his poem "Please Call Me by My True Names," excerpted here. He was moved by the true story of a little girl who drowned herself after being raped by a sea pirate:

> *Look deeply: I arrive in every second*
> *to be a bud on a spring branch,*
> *to be a tiny bird, with wings still fragile,*
> *learning to sing in my new nest,*
> *to be a caterpillar in the heart of a flower,*
> *to be a jewel hiding itself in a stone.*
>
> *I am the child in Uganda, all skin and bones,*
> *my legs as thin as bamboo sticks,*
> *and I am the arms merchant, selling deadly weapons*
> *   to Uganda.*
>
> *I am the twelve-year-old girl, refugee on a small boat,*
> *Who throws herself into the ocean after being raped by*
> *   a sea pirate,*
> *And I am the pirate, my heart not yet capable of seeing*
> *   and loving.*
>
> *Please call me by my true names,*
> *So I can wake up,*
> *and so the door of my heart can be left open,*
> *the door of compassion.*

Alan struggled to understand the balance of life while on Dancing Woman Mountain. He is far more "awake" than most people I know. His life experiences are extensive and he rises above despair, never abandoning hope. Hope is the miracle that "arrives every second," as Thich Nhat Hanh writes.

Seub's death was not in vain, nor forgotten. Nor did Alan's work go unnoticed. The younger generation in Thailand began to take matters into their own hands and bring the government to account for the decimation of the forests, the land, and the animals, which rightly belong to all Thai people. Seub was their hero, his picture carried in their hearts and on placards.

The royal family was moved by his death and martyrdom and allowed a statue of Seub to be placed in the park, the only statue erected in Thailand other than those of the royal family and religious figures. Corrupt officials were replaced and regulations in the parks began to be enforced. Conservation organizations around the world helped the young activists create fifty-one new national parks and wildlife sanctuaries between 1990 and 2000, bringing the total protected area to 16 percent of Thailand's landmass. The Huai Kha Khaeng Sanctuary is the last stronghold of the Indochinese Tiger in Thailand, except for a few on the borders of Burma and Cambodia. The park has been expanded, and a year after Seub's death it was designated a World Heritage Site.

4

Idaho

We were all playing hooky for ten days. It was a real vacation for five of us and a busman's holiday for the sixth. River rafting in the River of No Return Wilderness had been on our schedules for a year, and I was not going to miss out.

My staff at the National Endowment for the Arts was nervous. Committee appropriations were coming up for the agency and a senator would be irritated if he couldn't get the chairman on the phone. But I had been working overtime for eight months and this was a special vacation we had planned. I told my senior staff that I would return refreshed and ready to do battle once again, answering whatever questions Congress might have about controversial art and grants to our nation's artists.

Ed needed a break, too, from his grueling schedule as executive producer of television's *Law & Order.* Alan was between trips to Burma, since 1989 changed to Myanmar, where he witnessed ongoing poaching of Rhinos, Tigers, Otters, reptiles, and birds—a veritable cornucopia of wildlife. He was waiting for government permission to travel north to chart a vast, less accessible wilderness.

Alan was now married to Salisa, the beautiful geneticist we met at Dancing Woman Mountain. Dressed in a white and gold Thai embroidered gown, Salisa Sathapanawath was given away by my husband, her surrogate father, on the sloping lawn of our home in

Putnam County, New York, on a sylvan summer afternoon in 1992. We had encouraged the alliance from the beginning, telling Alan that it would take an unflappable Asian woman to live with a man of so many moods. They moved into a beautiful house on a ridge high above rolling hills not far from us, so we could hang out together drinking martinis and talking shop on weekends. Salisa received her master's degree from Mahidol University in Bangkok and got a job working in the lab at the Bronx Zoo under the renowned animal geneticist George Amato.

Alan's closest friend, Howard Quigley, a colleague from his University of Tennessee days, and his wife, Kathy, an Idaho veterinarian, made up the rest of our group. I was sure Howard and I were related—my birth name was also Quigley. Try as we might, however, we found no great-grandfathers in common, even though they all hailed from Ireland.

The twin-engine plane flew for an hour from the town of McCall over the Bitterroot Wilderness, the largest contiguous forest tract in the lower forty-eight states and protected forever by the visionary U.S. senator Frank Church, for whom a good portion of the wilderness is named. Our trip was a busman's holiday for Howard because he had long been studying Cougars in the Bitterroot as a project of the Hornocker Wildlife Institute. It didn't matter; he was happy to be trading cat scat for white water in one of the most beautiful wild places in America.

The plane bounced down on a grassy oasis surrounded by trees near the Selway River, and the director himself, Maurice Hornocker, stood waiting to greet us in front of a remote outpost of the institute. Maurice and George Schaller are the éminences grises of field biology today. Like George, Maurice traveled widely to study large mammals: Leopards in the Far East and Africa, Jaguars in Central and South America, and Siberian Tigers with Howard, who once gave me a plaster paw print of this largest of all the wild cats. It is the size of a dinner plate. Maurice is perhaps most famous for his work in the Americas on Canadian Lynx, Bobcats, Ocelots, and Cougars (also with Howard), and on Grizzly Bears, Wolverines, River Otters, and Badgers. He never leaves home without his camera and is as well known for his stunning animal photography as he is for his science.

He welcomed us into the low-slung log ranch house, decorated cozily with western blankets, pine furniture, and well-stocked bookshelves. The river burbled outside and yellow and black Townsend's Warblers flitted among the conifers. Deep in that forest in late June 1994 the tension of Washington, D.C., left me as quickly as water pops off a hot stove.

The oil lamp burned low at the dinner table as conversation about endangered species continued well past dessert. I looked across at Dr. Hornocker, movie-star handsome in the flickering light, and admired how he kept his positive outlook in the face of so much bad news. His wry sense of humor must have disarmed many a foe and won many a friend in his time. He considered the local people he'd worked with in Siberia or the Amazon equals, with as much innate knowledge of the land and animals as any college degree might con-

Left to right: Howard Quigley, Alan Rabinowitz, Ed Sherin,
and Maurice Hornocker

fer, and he decried the academic world, which demanded published papers and peer review before research was taken seriously. A naturalist such as the legendary Aldo Leopold would be given no quarter today. "Science asks you to publish three ways to determine it's dark outside," Maurice quipped.

Congress demanded similar credentials, and its reluctance to fund wildlife programs was of concern to all of us as we left the table and fell into our beds. I was having my own battles with Congress at the time as chairman of the arts endowment. The NEA, targeted by conservatives because of controversial grants, was the legacy I inherited when I took on the job. I made it my business to visit all fifty states to let the American people know what the NEA was doing for them in their district; I also met as many in Congress as possible. I was a staunch defender of freedom of expression and to the chagrin of the White House found myself in the newspapers more than any of us liked. The pressure was great from all sides: from artists who didn't think I was defending their work enough, to congressmen on both sides of the aisle who felt I had gone too far in defending those same artists, to the top brass in the executive office who didn't want me to embarrass the president. I had been paddling upstream for quite a while and welcomed this respite. A cool mountain breeze entered the dark room as we nestled under Hudson blankets and drifted off. I slept more soundly than I had in years. No phones, no roads, no people but us.

I dreamed of Sacajawea. The fourteen-year-old Shoshone girl, married to a French trapper, had endeared herself to Lewis and Clark through the Bitterroot Mountains trek on the expedition of 1804–06. She proved to be calm and resourceful in the face of crises, far more worthy than her husband, whom they had hired as their guide. Her baby was born in 1805, and she simply strapped him on her back and continued the tortuous climb through the crusty pinnacled peaks.

The Shoshone Indians had a three-thousand-year history in the area. Warring tribes pushed them higher into the mountains, where they survived on pine nuts, the occasional Bighorn Sheep, and the

Buffalo meat they dried on the Great Plains in the summer. Like most hunter-gatherers, they considered Mother Earth the source of all life and sustenance. She was to be respected and honored for the living things she provided. Human beings were only one among many and could take what they needed but must use what they took, as the wolf did the antelope, the eagle the salmon, and the bear the berries of the fields. The animals were accorded respect for the different attributes each had—the Coyote for its wiliness, the Cougar for its stealth. Everything had its own spirit, including the rocks, the rivers, and the tiniest insects.

This thinking changed as Native Americans began to be displaced by western expansion and the white man, with his ethic of dominance over the land and its creatures, and the concept of private property ownership. The most prevalent grazing animal on earth, the American Bison, or Buffalo, became scarce in the 1870s. The U.S. Army was responsible for slaughtering the most Bison, helping ranchers expand their cattle ranges and starving Indians in order to force them onto reservations. And warrior tribes like the Comanche killed as many as 280,000 Bison in one year to trade the skins, meat, bone, and horns.

The most plentiful large mammal in the world, which Plains Indians depended on for food and clothing, and which they revered as a spirit animal, was almost gone. Now the Indians had guns; buffalo jumps that herded the animals over cliffs gave way to bullets that dropped them where they browsed. Cowboys shot them by the hundreds to feed railroad workers, and also accompanied tourists and European royalty as they shot a Bison of their own for trophy.

The charismatic showman Buffalo Bill Cody killed a thousand all on his own. Forty million animals were reduced to one thousand by 1884. My grandfather Daniel Quigley was Buffalo Bill's doctor in North Platte, Nebraska, where Cody lived with his family in a fine old Victorian house. The Wild West show rehearsed staged Buffalo hunts and Indian wars in North Platte before taking the famous show worldwide. My grandfather told us that Cody deeply regretted his part in the decimation of the Bison, and that he fully supported

President Teddy Roosevelt and the New York Zoological Society in founding the American Bison Society. In 1901 the zoo's director, William Hornaday, imported a small herd of Bison to a meadow near the Bronx. From this breeding group came almost all the subsequent Bison that repopulated the West, saving the species from extinction.

This was the beginning of the conservation movement. Teddy Roosevelt, a renowned hunter, was our first and greatest conservation president. No other administration in U.S. history has placed conservation of the environment as a top priority for the federal government. Although TR believed in logging the forests, in damming the rivers, and in creating parks for people as much as for animals, and often disagreed with purist preservationist friends like John Muir, who founded the Sierra Club, Roosevelt's record was extraordinary. He established five national parks, four game refuges, fifty-one bird reserves, and one hundred fifty national forests during his presidency.

Conservation is an attitude, a spiritual belief, or a regulation, and for the first two hundred years of American settlement it didn't exist. We were a rural society, a society that believed in taming the land and putting the plow to the earth. Wolves, Bears, Coyotes, and Cougars—anything big, competitive, or dangerous—were to be exterminated. Wolves were all but extinct in the lower forty-eight states by 1960, and Grizzlies in all but a few pockets. Coyotes became cleverer, managing to thrive despite mass poisoning, trapping, and shooting. Cougars, extirpated in the East except for the Florida Panther, hung on in the West despite efforts to eliminate them. Their secretive ways, their stealth, which the Shoshone saw as a holy inner spirit, kept them alive. The life of the hunter-gatherer, who depended on natural resources being available year after year and prayed to Mother Earth for bounty, was over, fallen to the cultivators of the earth who prayed to God for their crops to grow.

In 1994 Maurice Hornocker still felt hopeful about the future of large mammals in North America. He said that as the United States urbanized in the mid-twentieth century, the desire to protect the wilderness and the creatures that lived in it began to grow. By 2014 fully 50 percent of the population lived in cities, and that number may increase to 80 percent by 2100. It is easy for those who don't confront poisonous snakes or big cats on their city streets or in their backyards to take on the cause of conservation, but Maurice pointed out that there were changing attitudes among rural people as well. Some ranchers were promoting the health of the land through a balanced ecosystem that maintains apex predators like the Cougar. Maurice was excited about the reintroduction of Wolves to Yellowstone and elsewhere. The people of California made it illegal to hunt Cougars in 1990, despite incidents of mauling and killing of pets. Maurice believes there may be more Cougars in America today than at any time in their history. He felt there was much to be done but that it was all possible with social and political will. Social and political will—getting the support of the people and the politicians who can enact and enforce legislation—is the crux of the matter. There in Bitterroot Wilderness the solutions were clear. Getting there was another thing.

On the river I was content not to paddle at all. I lay in the stern of the raft and gazed at the treetops and sky, watching the Osprey or the Harlequin Ducks, or the mammoth trout swishing in the eddies below. When the roar of the river increased and a flurry of white water tossed us about, I held on for dear life, sometimes flipping in the air like a child in a bouncy castle. At one point Ed and Alan, twelve feet ahead of me in the bow, became totally submerged as we raced down a waterfall, and then appeared again paddling like hell to bring us up again. Those minutes were exhilarating. Alan whooped with delight. I had never seen him so carefree and happy, his new bride by his side and surrounded by the love of friends.

Most of the time we just drifted downstream, quiet in our own thoughts, connecting with nature. At night we camped under the stars and warmed our cold toes by the fire, gloriously content in this protected wilderness, the regulation of which was so strict we had to bag our feces and urine to take out.

Howard's work was to preserve habitat like this from ever being

Rafting the Selway River, Idaho, 1994

See the flag? The raft is under the white water!

developed so that the Cougar, the Badger, the Black Bear, and the Golden Eagle still had the space they needed to survive. Alan was about to start his work in Burma assessing the Tiger population there and ways to keep it from extinction, and I was going back to a hostile Congress seeking to gut the NEA.

The idyll on the river wound down. We felt good despite the uncertain future of our causes. We were burnished by the sun and blissfully satiated with the heady effects of the great outdoors. It took more than an hour on the single dirt road to reach a lone general store on the edge of the River of No Return Wilderness; a sign said, "Lowell, Idaho, population 23." As I stepped from the van a woman hurried toward me from the store. "Are you Jane Alexander?" she said. "Call your office immediately." End of idyll.

5

Nepal

The tiny window of the plane could not contain the mountain—not even half of it. I scrunched down in my seat, craning my neck, and there, as close as I'd ever get, were the tops of the world's highest peaks riding the blue like the white sails of a sky goddess. Terraced rice paddies and fields of winter wheat climbed the slopes to the tree line. A huge valley came into view and we descended toward the dusty city of Kathmandu, gateway to the Himalaya.

We spent two days in Kathmandu acclimating to the altitude. Nepal's capital was going through a revolution in 1990. The king's men, khaki-clad soldier boys, were patrolling the streets, and an occasional pop of rifle shot could be heard in the alleyways. The People's Movement, which by 2006 demanded the complete over-throw of the monarchy, was just beginning but did not seem to disrupt the comings and goings of trekkers, soul searchers, potheads, and biologists. It was a funky city, bustling with hawkers of trinkets and hash pipes, the Hindu temples nestled near Buddhist prayer wheels in the confusion that accompanies overlapping cultures. Poverty was etched on the people's worn faces, their dress, and the collapsing balconies of the ancient architecture. Men, women, and children pursued us in the streets, desperate to sell Gurkha knives, strings of beads, or cigarettes. A Snow Leopard pelt was selling for a mere fifty dollars in the back room of a little shop.

A Snow Leopard in Nepal, taken by the wildlife photographer Steve Winter

The Snow Leopard is the iconic cat of the Himalaya, both revered and hunted. It roams the high mountains from India to China, and from Afghanistan to Burma. No one knows how many of these beautiful animals are left. They are secretive and hard to see in the rocky, snowy terrain. When they cannot find Blue Sheep to eat they come down and raid farmers' livestock. Meat is laced with poison and the cats die.

Rodney Jackson was in Kathmandu while we were there. This was his tenth year of radio-tracking Snow Leopards in Nepal, and it would be another decade before his Snow Leopard Conservancy, with its mission of harmony between man and cat, was founded. Few scientists ever set out to study the Snow Leopard with his steadfastness. Rodney and his partner, Darla Hillard, worked with hundreds of local farmers to find ways for them to coexist with the Snow Leopard and become its guardian. They instituted a community-controlled insurance policy for livestock, where the farmer pays a small premium up front in return for a sure price for his domestic animal should the Leopard take one. Although the chance of seeing a Snow Leopard is slim, many Westerners are signing up for tourist visas to the area, an additional incentive for the local people to keep the cats alive.

George Schaller continues to return to the vast Himalayan plateau he so loves. He encouraged Buddhist monks in Tibet and elsewhere in the Himalaya to keep the hundreds of monasteries that dot the landscape sacred and off-limits to hunters. He wrote that if the Snow Leopard was not protected, the mountains would become "stones of silence"—the title of his classic book on the great cat.

Alan, Salisa, Ed, and I began our five-day trek of the lower Annapurna range from Pokhara in the western part of Nepal. We set off with Sherpas and a ridiculous number of porters who carried everything we had and all we might need for tenting and meals. At least we were employing a good number of them. They leapt ahead of us with boxes, bags, pots, and pans hanging down their backs from tamp lines taut on their foreheads. Some were barefoot, preferring the security of splayed toes and the soles of their feet to the soles of their shoes. I marveled at their agility on rocky inclines. Alan kept

*Salisa and Alan Rabinowitz and I picnicking beneath
the Annapurna range, Nepal, 1990*

pace with us for a few miles and then, in a burst of macho flair, grabbed a duffel from one of the porters, held it on his head, and raced forward. We didn't see him again until the end of the day. This was typical of Alan; he is hard to keep up with, literally and figuratively, accomplishing more in a week than most of us do in a month. Salisa is lithe and a runner; she kept pace in the middle of the pack, while Ed and I brought up the rear.

I was drawn to the lush flora covering the hillsides and the birds among it. Glossy Ashy Drongos perched on long branches in the morning sun; Indian Rollers undulated by, their turquoise and teal-blue wings unmistakable in flight; while the stunning scarlet of the Long-tailed Minivet brightened the green foliage. The Great Himalayan Barbet called loudly in the oak forest, and various species of vultures, kites, and eagles rode the thermals high above. The beauty was staggering because the mountains, peeking through the clouds so far above us, made everything seem part of some mythical kingdom.

Surprisingly, we met no other trekkers, but we passed through many villages of different Hindu caste peoples, first the Chhetri and later the Gurung. Some women came and sat with us as we ate our lunch on the ancient stone paths. They sat on the ground and stared at Salisa and me. They were lovely with small rings in their noses and ears, and silver necklaces draped around their necks. I wondered what they were thinking. They did not ask for anything, not money or even a smile; they just stared. Their families had lived there for hundreds, maybe thousands, of years, and their red adobe-like homes, rectangular or round with a thatched or slate roof, were beautiful to look at, reflecting the order of centuries. Soon they would be harvesting their crop of winter wheat.

The eastern Himalaya region is known for its diversity of plants, with multitudes of orchids and primulas, and as the hub for genus diversity of rhododendrons in particular. There are hundreds of species. I have never been in these mountains in April and May when the riot of color is at its height, but in March a few species of red flowering rhodies were spotting the slopes and little sunbirds were reaping the nectar while pollinating the plant at the same time.

This trek was not difficult. We stayed below twelve thousand feet and followed well-worn trails. The tenting at night was on fields laid bare by the many who had come before us. Ed and Alan dug individual pit holes for each of us and were hard put to shovel unearthed ground. Children found us each evening as the snowy peaks of Annapurna turned golden in the waning light; they laughed and tested their English.

I saw no wild animals except birds and insects during our time in Nepal, but we were in areas long civilized. With more than thirty million people living in a country the size of Arkansas, Nepal manages its wild habitats, a healthy 29 percent of the country; everything is cultivated or managed except for the highest mountain peaks. There are twenty-three parks and reserves, including those where hunting is allowed. The country is a model in Asia for the administration of its parks, two of which are World Heritage Sites: Chitwan in the south, and Sagarmatha, the premier mountain park of the world, with Everest at its heart, in northeastern Nepal. Sagarmatha is an IBA—an Important Bird Area—with more than one hundred species, high-altitude beauties like the Blood Pheasant. Its mammals include the Musk Deer and the Himalayan Black Bear, as well as the Snow Leopard and the Red Panda.

Chitwan, the oldest park, has had some troubling times. It is one of the last places to see the Indian, or One-horned, Rhinoceros, whose population wavers depending on the success of poachers, who continue to kill these ancient beings for their horns. The going price can be higher than $100,000 for a single horn, used by the Chinese and especially the Vietnamese for spurious medicinal purposes. The Rhino population plummeted from eight hundred animals in the 1950s to three hundred by the end of the century. In 2003 alone, thirty-seven Rhinos were killed in one massacre. The government of Nepal ordered soldiers into the park to handle the situation, and by 2011 no Rhinos were being killed by poachers, a clean record still maintained by Nepal. The soldiers, heavily armed, patrol by jeep and on Elephant where they can move through the thick grass in pursuit of poachers. By 2012, the Rhino population had rebounded to five hundred animals, but the poachers have become more sophisticated in their methods and are backed by international cartels of criminals. Wildlife trafficking is so lucrative that it ranks fifth in international illicit trade, after counterfeiting, drugs, the trafficking of people, and illicit oil.

It is harder to estimate the number of Bengal Tigers killed by poachers in Chitwan because the entire body of the animal is taken; the skin, the head, and the bones are all valued in illegal trade. The census of 2013 put the number alive in the park at about 125 breeding females, a figure the government wants to double by 2022, the next Year of the Tiger. With an estimate of only one thousand wild breeding females left on earth, Chitwan's goal is admirable.

Nepal's managed parks and reserves are really mega-zoos, a popular solution to saving large mammals globally. Dr. William Conway was prescient in predicting this in the 1980s when he claimed that with a rising human population of ten billion by the mid-twenty-first century there would be no way to safeguard wild animals except in isolated parks or "mega-zoos." By the 1990s Dr. Conway insisted that zoos and all field biologists needed to be dedicated to conservation, not just research. It was in 1993, under his direction, that the esteemed ninety-eight-year-old New York Zoological Society changed its name (and with it, its mission) to the Wildlife Conservation Society.

Nepal's leaders have halted poaching by bringing in the army, while juggling the needs of thirty million people who speak a multitude of languages and whose cultures often clash. The government recognizes the value of the country's wildlife, especially as it relates to the lucrative tourist industry. They are taking a hard line against criminals in wildlife trafficking and have committed manpower to get the job done. But with the devastating earthquake of 2015, resources will be needed for decades to come for Nepal's displaced people first and foremost. The needs of wildlife may have to be sacrificed.

6

Myanmar

The 1990s were about politics—for Alan, for me, and, it seemed, for all the nongovernmental organizations involved with conservation and nature. The Internet was nascent in 1995, but within ten years it would demonstrate how online communities could put pressure on elected officials to enact legislation. It still had a long way to go when I was in D.C. As President Bill Clinton said years later, "When I took office, only high-energy physicists had ever heard of what is called the World Wide Web . . . Now even my cat has its own page."

I sat in my office at the NEA, writing some remarks on my electric typewriter, when the three techno-devotees on my staff of three hundred came in and urged me to get "wired." Vice President Al Gore had requested as part of his "reinventing government" dictum that all federal agencies begin communicating internally and externally on the World Wide Web through computers. I went to visit him at his White House office to talk about the NEA's budget and problems I was having with Congress, and he ushered me excitedly to his desk to see a special website pop up on the screen. He saw the future of the Internet long before anyone else I knew. He also assured me that the White House was behind me in my struggles.

The techies wired me first with a fine huge Macintosh; they taught me how to negotiate the complex systems of the machine and how to surf the Web. The world opened up in ways I never dreamed of as a child. I could access museum exhibits, photographic archives, and the early digital works of online artists. The Web compiled the latest migratory bird patterns and gave me all the works of Shakespeare at the click of a button, which truly thrilled me. Although government bureaucracy and ineptitude kept the NEA from being fully wired internally, the Web bunch created a great site for us, plugging in audios of blues singers and string ensembles to whom we had awarded grants and listing all essential information about the agency. We did it in record time, and Al was proud. The Internet is a miracle, and nothing will ever be the same. Fully 40 percent of people on the planet today, more than three billion of us, are online.

While I was dipping into the future of communications, Alan was going back into the past—way back. Between 1993 and 1999 Alan journeyed to many remote areas of Myanmar, formerly known as Burma. As director of Asia programs for the Wildlife Conservation Society he dreamed of going where few had ever gone and to a country that had been closed to foreigners for three decades. He stared at the huge expanse of wilderness on the maps conjuring Indochinese Tigers, Asian Elephants, and Sumatran Rhinos. When permission to travel finally came, it was from the military government, the State Law and Order Restoration Council, or SLORC, one of the world's most flagrant violators of human rights, routinely perpetrating mass murder, rape, torture, recruitment of child soldiers, and forced labor. Aung San Suu Kyi should have been the head of government after elections but was instead under house arrest, where she remained until 2010. The military had taken over all departments, including forestry, in the name of "national security," and were exploiting natural resources to the point of unsustainability. Only 1 percent of the land was protected in parks, and nothing was truly protected as there was no onsite staff to manage poaching. This was the situation when Alan arrived in Yangon (formerly Rangoon) in 1993.

Many cautioned Alan about working with a despotic government, and some felt that it was morally wrong to do so. But animals don't engage in politics—except those of the human persuasion—and most have been on earth for millions of years, not the mere two hundred thousand we humans have. I encouraged Alan to do what he could with whomever he encountered. My brief time in Washington had shown me that politics does indeed breed strange bedfellows. Who would have guessed that one of my staunchest allies, a lover of the arts, would be conservative Republican Orrin Hatch from Utah?

Alan was in luck. SLORC cabinet member General Chit Swe took great interest in the wildlife of his country and in preserving the forests. He endorsed Alan and WCS as advisers to the government, and Alan was cleared to travel with forestry staff in secure regions of the country.

Alan was interested in knowing the status of the Sumatran Rhinoceros. With George Schaller, his mentor and the director of the WCS Science Program, he traveled the Chindwin River north to the largest protected area, the Tamanthi Wildlife Sanctuary. No Westerners had been there for decades, and they expected that these old hunting grounds for British sportsmen would reveal numerous large mammals including Indochinese Tigers, Asian Elephants, and the endangered Sumatran Rhino, which was barely holding on in Sumatra and was extinct elsewhere.

The small Sumatran Rhino probably looks today as it did twenty million years ago when it shared eastern Asia with three-toed horses, early apes, and thriving whale species in the kelp oceans. It survived millions of years of the planet warming and cooling, and the Himalaya shifting higher and higher as tectonic plates pushed against each other. The ignominious demise of the Sumatran Rhino came about because men coveted the two horns protruding from its snout. Through the ages the keratin horn has been consumed as a cure-all for poison, cancer, impotence, headache, fever, and just about anything else you could name. You might as well grind up your fingernails or hair and eat them for all the validity of the claims of traditional Asian medicine. But the superstition has persisted for thousands of years and targets every Rhino living today with a death sentence, because the horns are worth their weight in gold.

Alan and George had some hope that in this remote part of the world the Sumatran Rhino had managed to elude hunters and survive. What they found was a beautiful high-canopied forest devoid not only of human life but also of large animals. It didn't take long to understand why. They were camped by the water one evening, the soldiers accompanying them in civilian dress, when three hunters calmly walked in, thinking they were locals. Their packs were filled with poaching snares and the body parts of otters. They were confused and surprised at being arrested; they had hunted this area for many years and no one had ever told them it was wrong. They sold Tiger and otter parts, bear gallbladders and Rhino horns across the border in China, but they said that all the Rhinos were gone now. They were still getting a Tiger a year, and showed Alan how they set up the snares with bamboo spikes to pierce the Tiger's flesh and weaken him.

The forestry department responded to Alan's assessment of the sanctuary by building a ranger station and educating the local people about poaching. The Wild Bird and Animals Protection Act had been in place since 1912, but Alan doubted that the trade in body parts could be stemmed. The culture of killing was growing and spreading like a cancer all across Asia. At least in Myanmar he didn't seem to be dealing with government corruption as he had in Thailand. It's anomalous that dictatorships often make the environment safest for animals. In Myanmar, Burmese soldiers had a long history of discipline under British rule; corruption was as unacceptable to them as it was at Buckingham Palace.

Alan did an assessment for General Chit Swe of a pristine island called Lampi, where the local people had decimated coastal species such as sea turtles, urchins, and cucumbers for the Thai market. Most of the large trees had been cut down by timber poachers and fish were being blasted out of the water with dynamite. Still, Alan saw promise in species recovery if part of the island was made a protected area and therefore off-limits. The general wasted no time in seeing it designated Myanmar's first marine park.

Alan kept staring at the map. Farther north, near the borders of India, western China, and Tibet, was a transition zone, where the flora and fauna of lowland tropical species overlapped with upland Himalayan species. This would be where the Pangolin, the Jackal, and the Palm Civet met the Asiatic Black Bear, the Red Panda, and the Goral. After the interminable and obligatory rounds of government officials Alan finally set out with a hundred porters, an orchid specialist, an ornithologist, a medical officer, and the Burmese biologist Saw Tun Khaing, whom he had hired to coordinate WCS's activities. This was the biggest trip Alan had ever taken and he was to be gone for two months, a long separation from Salisa back in New York. He needed to know what was hidden in the depths of the vast north.

In Putao, the local market and almost every household revealed animal skins and parts that people were eager to show him. When word got around that the Westerner was interested in animals, hunters came and opened sacks filled with deer heads, turtle shells, hornbills, monkeys, and Leopard and other big-cat skins. They said they didn't find Tigers anymore. They dried their catch and waited for traders from China. They talked freely, having no idea there were hunting laws.

It disturbed Alan that he had to buy the animal parts; it made him as guilty of the trade as any Chinese merchant. But he justified it because there was no reference collection of the species anywhere in the country and he had promised the government he would help with a biodiversity exhibition. He also needed to study the animals in detail.

Still, I thought, was it necessary any longer to be collecting species at all, as museums across the world were still doing? There seemed to be no coherent policy, since each museum sought its own collection and paid little heed to the killing of threatened species in the last strongholds on earth. Nor did they seem to investigate fully whether another institution might have the animal already. I once asked to see the presumed extinct Ivory-billed Woodpecker in the drawers of the American Museum of Natural History and was stunned to learn that they had *thirty* of them squirreled away. There are more than four hundred of them in museum collections. (Oh, so *that's* why they went extinct!) For all their scientific value, I respond to Thoreau's description of these bastions as "the catacombs of nature."

If it hadn't been for the hunters Alan probably would not have discovered a species new to science. A little twenty-five-pound deer with orange fur was placed in front of him one day and he immediately sensed that it was different. He bought the carcass, and many months later, back at the Bronx Zoo, George Amato determined the DNA and substantiated Alan's suspicion: this Leaf Deer might have been known to the people of northern Myanmar, but it was new to the world of science.

As the team continued its journey north, past the transition zone and into the Himalayan region, the party left behind the leeches that made their feet and ankles bloody, and the biting insects that welted their bodies. They began shivering in the cold mountain air. The sheer number of animals they saw in huts and in hunter's sacks attested to remaining populations in the wild, even if the Tiger, Elephant, Takin, Serow, and Red Goral seemed headed for extinction like the Sumatran Rhino. Alan began to envision it all as a protected area with hunting regulations and strict enforcement. The animals would thrive if given half a chance, just as the Jaguars had in Belize's Jaguar reserve.

The trade in animal parts in northern Myanmar had been going on for as long as anyone could remember. The most coveted commodity being traded for animal parts was a complete surprise. It was not exactly greed that was driving the trade, but necessity. This part of the world lacked a vital mineral: salt. Alan said it was like going back in time thousands of years. Roman soldiers had been given salt for their work; it was called their "salarium," or salary. Mark Kurlansky wrote an entire book, *Salt,* about the history and value of the mineral.

Human beings need less than a gram of salt daily to keep the nervous system working properly. Most of the salt we need is ingested in the food we eat. Because the Myanmar staple is rice, which is naturally low in sodium, and the northern Myanmar diet adds meat only sparingly to dishes, it is possible that the people Alan encountered were truly sodium deprived. Whether the northern Burmese actually needed additional salt or whether they simply craved it, as human beings and many animals do, the trade existed and was decimating wildlife. If the area became a park and the people were supplied salt in exchange for not hunting endangered animals, might that be a solution?

As the weeks went by and the team trekked higher into the lush mountain valleys, everyone suffered from exhaustion and physical ailments. The stress of talking to every village about the animals they hunted took its toll. One of Alan's knees was killing him, and he spent sleepless nights when his thoughts turned constantly to Salisa. The porters were mostly teenagers, and despite the toughness of the journey they told Alan they were having a great time. The boys were getting to see a world beyond Putao, and the girls laughed, saying they were happy to be away from their husbands.

In northern Myanmar, Christian missionaries had converted most of the people. Still, the awesome power of the young Himalayan Mountains inching ever skyward, the rivers tumbling from them, and the lush green valleys failed to quell an innate animism. In his book *Beyond the Last Village,* Alan wrote:

> I always feel both strengthened and humbled by the almost palpable energy of truly wild places such as this one. It never surprises me that human beings feel a need to worship and appease the power of raging rivers and the indomitable presence of towering mountains. To me it is a manifestation of the intuitive understanding that one has to live with nature, instead of constantly pitting oneself against it, in order to survive. What saddens me, however, is when people learn to suppress such feelings, becoming convinced that humans are apart from nature.

The Christian interpretation of man's dominance over animals refers to Genesis 1:28 in the King James version of the Bible:

"Replenish the earth, and subdue it: and have dominion over the fish of the sea, and over the fowl of the air, and over every living thing that moveth upon the earth."

Alan met Christian preachers in many countries, including Myanmar, who took this biblical quote as a license to kill indiscriminately. More than once he argued with them and their followers, especially when they took pregnant or lactating females and young. But in these Himalayan villages there were also Christian leaders who were wise, who looked instead to the verse from Isaiah 11:6: "The wolf also shall dwell with the lamb, and the leopard shall lie down with the kid; and the calf and the young lion and the fatling together; and a little child shall lead them."

Alan made more trips to Myanmar in the late nineties. He and Salisa trekked new routes in the north where he talked about a future when the villagers would have salt and other needs met and animals could thrive. He made friends with the final generation of a tribe of pygmies known as the Taron, twelve little people who had decided not to have children because of inbreeding. He witnessed these human beings make a conscious decision to go extinct. Perhaps they had been caught, like so many of the unique mammalian

Alan Rabinowitz with one of the Taron pygmies headed
for extinction, in the remote mountains of Myanmar, 1990s

species in these remote mountain outposts, many thousands of years ago with nowhere to go and no escape.

Khaing and Alan drew up plans for protected areas, larger than any already existing in Myanmar because the big animals needed space to become viable populations if they were to survive. They presented the plans to the government, only to learn that their supporter General Chit Swe was gone, perhaps under house arrest. No one seemed to know what was happening as the government reorganized once again. Alan waited for a response as he and Khaing educated villagers about animals and how to protect them from poachers. Khaing was eager and ready to take over the responsibility of the leadership role. He had been by Alan's side for most of the 1990s, through hardship and hope. He learned from Alan's passion despite their differences. He often told Alan to learn to accept the things he could not change. Alan would reply that he had to *try* to change what he could and that he "reserved the right to rage against even those things I can't change." Besides, he added, "It reminds me that I'm still alive."

He returned to his country home in New York to be with Salisa for the coming birth of their first child. He went to the Bronx Zoo and ran the Asia program while waiting for news from Myanmar. On the cusp of the new millennium, November 4, 1999, Salisa gave birth to a beautiful baby boy with the dark hair and eyes of his mother. Alexander Tyler Rabinowitz exhibited the fortitude of his father from the very beginning of his life.

Finally, on April 3, 2001, word came that Myanmar's ministry of forestry had created the protected area Alan had proposed, the Hukaung Valley Wildlife Sanctuary. This, along with the Hkakabo Razi preserve authorized earlier, and almost double the size Alan suggested, made this the largest protected wildlife area in all of Asia.

7

India

It was not possible for me to travel to Myanmar in the years Alan
was there. The military government probably would not have issued
me a permit even if I had been able to go. Besides, I was in Washington fighting to keep the NEA alive. In 1995 the first Republican
Congress in forty years was elected, with Newt Gingrich as speaker
of the House. Their "Contract with America" included the complete
elimination of the NEA, as well as the National Endowment for the
Humanities, the Public Broadcasting System, and the Pell education grants. I went on the road, bringing my case to the people. I
traveled to all fifty states, and districts such as Puerto Rico, meeting and talking to people about the arts the NEA helped fund in
their neighborhoods—everything from craft and music festivals to
museums, dance, theater, and war memorials. It was not a hard sell.
People love the arts: it gives their community identity, entertainment, and a spark of imagination. Once they connected the NEA to
the places they frequented, they didn't want to lose the endowment
and the grants it gave. Most people didn't care about the controversial grants that disgusted some members of Congress and forced us
into the headlines too often. As long as people were told beforehand
what to expect of an exhibit, they were content to make their own
choices. It was a matter of education. In the end the people saved
the agency by pressuring their legislators to vote for it to continue.

Senator Alan Simpson was one of the few Republicans who supported me, the First Amendment, and the NEA through thick and thin. And we had Buffalo Bill in common: Al's grandfather was Buffalo Bill's lawyer, and my grandfather was his doctor. Here we are celebrating at a fund-raiser for the Buffalo Bill Museum in Cody, Wyoming, in the 1990s.

But it was a tough few years in the trenches of politics before the battle was over.

Alan Rabinowitz and I would compare notes on weekends in our kitchen, when he was home from his mountain battles. You couldn't second-guess politicians; they had their own agenda and sometimes, like the conservatives we were both dealing with, they would do an about-face and surprise you. We were both high from our victories but weary as well.

Alan was physically weary. He is one of the strongest men I have ever known and keeps fit through daily weight lifting, boxing, and wrestling. His muscles are taut, and an invitation to punch his abdomen is more likely to hurt the puncher's fist than Alan's gut. His global travels take an enormous toll. He has had parasites in his system for months on end, broken bones, infections, and injured knees that would not heal, but he always recovered and went back out to face the newest challenge in the wild. If he was taking longer than usual to recover, it was simply a matter of time before he was his old self again.

He was not prepared for the diagnosis his doctor gave him just two years after Alex was born. Chronic lymphatic leukemia is an incurable cancer. Alan was told he had four to twelve years to live. He was forty-eight, and Salisa was pregnant again.

Alan was remarkably composed for a man living with a death sentence, but he had escaped death numerous times already—plane crashes, truck accidents, mountain ledges, fierce animals, and men with guns. Perhaps he thought he'd beat the odds, and if not, it had been a pretty great ride.

Tears clouded his eyes when Alana Jane Rabinowitz was born on May 23, 2002. Ed and I, the children's godparents, were present at her birth, as I had been at Alexander's as an unofficial doula. Salisa's slim hips closely cradled her baby girl, and her labor was long and hard. Alana Jane finally arrived, beautiful like her mother, with a head of lustrous black hair and deep brown eyes. She began to wail. She did not want to be washed or wrapped or fussed with. Alan stroked an exhausted Salisa as Ed took the baby into his arms. He gently rocked her, walking around the sterile hospital room and toward the light streaming through the window. Alana quieted instantly and gazed at the bright new world she had become part of.

Alan's leukemia did not bring him down. He had too much to live for and too much to do to let the cancer stop him. His doctors at Memorial Sloan Kettering in New York gave him a cocktail of drugs to combat the disease, although there was no known cure. Hedging his bets, Alan also consulted a Chinese shaman in the heart of Chinatown. He sat in a tiny room facing an intense little man who spoke no English, who stared at Alan, transferring an energy that made Alan feel better after each session.

His fiftieth birthday was celebrated at a rousing party given by his sisters, who jumped the gun by a year. When your birthday is December 31 you can get mixed up. Alan was really only forty-nine, but it didn't matter, we needed to celebrate his life. He was feeling okay and was ready for travel.

As head of the Asia program, Alan was going with George Schaller, other staff, and us trustees to observe the Bengal Tiger programs that WCS funded in India. India has forty-two Tiger reserves and the most Tigers in the world. The historic range of Tigers extended through eastern China and Malaysia south to Indonesia, and north and west through Russia and Turkey. Today they live in only 7 percent of their former range. All subspecies, from the Amur (or Siberian) Tiger in Russia, with only 400 animals, to the South China Tiger, which hasn't been seen in the wild in China for ten years, to the Sumatran, with about 400, are being pushed into smaller and smaller areas as the human population advances. Wild Tigers today, optimistically estimated at about 3,200, are highly endangered.

There is probably no animal on earth so heralded, feared, and revered as the Tiger. Their veneration is exceeded only by their slaughter. Tigers preceded *Homo sapiens* on earth by as much as a million years, but once we made our way out of Africa the need to triumph over the top predator of Asia must have been overwhelming. Like two heavyweight champions in the ring, we have been circling each other throughout time. Tigers were captured and pitted against Lions in Rome's Colosseum, the Asian outlier against the North African regular. The only report of a battle between these two greats—in 80 AD, by the historian Martial—said the Tiger won. The Romans decimated much wildlife with their bloodthirsty shows, but it is doubtful that they made much of a dent in the Tiger population, because the Tiger's range was so great. Rome's use of mammals, including human beings, as commodities for entertainment paved the way for sport hunting by India's Mughals, who ruled from 1526 to 1857. The Mughals, intent on bringing down the Tiger with sword or arrow, subsumed the divine power of the animal by killing it. The Tiger might escape the sword or the arrow, but when guns took over the hunt, and beaters drove the animals from their lairs, they had no chance.

The massacre of Tigers in the late 1800s was not unlike that of Bison in North America at the same time. The maharajas prided themselves on showing their European guests a good time, especially their British masters. They increased their own status with each Tiger they killed. One maharaja boasted that he shot 1,100 of the great cats. There are photographs of the beautiful animals lined up on the ground like so many logs, their killers behind them as proud as if they had exterminated vermin, which is, indeed, how the cats were viewed. Between 1875 and 1925 it is estimated that 80,000 Tigers were taken, although no one knows the exact number. Journals complain of fewer cats as the years went by. With India's independence in 1947, trophy hunting became the rage. The ordinary man in the new democracy, as well as sportsmen from the United States and Europe, could each have their Tiger. High-powered rifles, scopes, and jeeps ensured them that they would.

Then things started to change. Indira Gandhi became prime minister in 1966. She was a huge champion of her country's wildlife and was alarmed to learn that the Tiger population had plummeted from 40,000 in 1900 to just 1,800 in the 1960s. Soon legislation imposed strict hunting laws and it became illegal to export Tiger skins. She appointed a task force in 1971 to study the situation, and Project Tiger was launched in 1973. Nine Tiger reserves were established, guards were brought in to patrol, and entire villages were moved. Indira Gandhi understood the whole complex interaction of Tigers and men and the need for healthy habitats for the species' survival. The numbers of Tigers more than doubled, to 4,000, in the ten years that followed. Then, in 1984, Indira Gandhi was assassinated by her own Sikh bodyguards and the brief bright light of the decade was dimmed again, never to fully recover.

Dawn in Kanha was golden. Rays of sun crept across the pastures and illuminated the Sambar Deer, its majestic antlered head alert to our approach, and the smaller Swamp Deer bounding from the pond's edge over the fields. These prey animals keep the Tiger alive. The Swamp Deer, or Barasingha, is found only here and two other places in the world, but it shares Kanha's lush swampy lowland with the little Spotted Deer (or Chital), the Muntjac (or Barking Deer), and the Wild Boar. The ancient villages that once dotted this area are all gone and their cultivation has now reverted to mixed grass-

Kanha at dawn with the Tiger's favorite prey—grazing Spotted Deer

lands edged with forest. It is one of the most beautiful of the Tiger reserves.

Our birding team of four, plus Raj Singh, the naturalist and bird guide author, was up and at 'em before all the rest, as is usual with birders. We go slower. Those who are mainly interested in seeing mammals are hot to race around and find them. We amble. We search the trees and the sky, as well as the ground. We take our time, but we miss very little, and at the end of the day we invariably have as many mammals, if not more, than the others, in addition to a hundred birds. This morning was no exception.

India is blessed with abundant bird species. In the crush of humanity there is still respect for life. Royalty may have their hunting preserves, but the culture of killing was never really the domain of the everyman, who saw life as sacred. White cows walk the thoroughfares with the equanimity of a maharani, and Macaques steal your lunch or your rings and are not poisoned. Birds ride the skies with carefree abandon, more than a thousand different kinds, multicolored and exotic.

In Kanha the wild Peacocks were screaming and strutting their glorious tail display through the foggy clearings as a reminder that the world in all its imagination held nothing more beautiful than they. A pair of Collared Scops Owls peeked out of their nest hole, and a Painted Partridge, speckled black and white with a dash of russet like a helmet on its head, scurried across the dusty dirt road. You can get jaded in India as one stunning bird after another crosses your lens: Crested Hawk-eagle, Sirkeer Cuckoo, Brainfever Bird (which monotonously calls "brain fe-ver, brain fe-ver, brain fe-ver" until you get the disease), Green Bee-eater, Grey Hornbill, Brown-headed Barbet, Large Cuckooshrike, Black Stork, Coucal, Red-rumped Swallow, Black-headed Oriole, Scarlet Minivet, Plum-headed Parakeet, Paradise Flycatcher with its long white tail, and on and on and on. Just saying the names is exotic.

By 10 a.m. we were on the backs of Elephants crashing through a dense forest of sal and bamboo in search of a Tiger that was eating his kill of Spotted Deer. We went in one and two Elephants at a time to where the Tiger lay in the brush, so well camouflaged in the flickering sunlight that had you been walking ten feet away you would never have seen him. He growled at the intrusion of his mortal enemy the Elephant but stayed protectively near his bloody kill. He was sleepy with satiation, which gave the four of us sitting ten feet above in the howdah a perfect opportunity to gaze at his magnificent coloration. No two Tigers are alike. Of course, no two *anything* in nature are alike, but a Tiger's stripes and dots are clearly identifiable. This was a magnificent young male, his pale yellow eyes fixed on ours until I felt sure he would rise from his morning bed and leap on the Elephant's thighs toward us. I was sorry we had disturbed his quiet refuge, but I was not sorry for the indelible image of his magnificence, which I carry with me to this day. I had looked into the eye of the Tiger.

> *Tiger, tiger, burning bright*
> *In the forests of the night,*
> *What immortal hand or eye*
> *Could frame thy fearful symmetry?*

So begins William Blake's seminal poem. It is not a poem about seeking to know the Tiger but to know its creator. It is a poem of wonder, of the inexplicable, of the mystery of being. In an age of reason, how do you explain the Tiger? It is also a metaphor for our own divided nature: "Did He who made the lamb make thee?" As human beings, we long for the gentler, rational side, our better natures, but in reality we must confront the ferocity of the Tiger in us, the wild part. We are apex predators alike, and the predator in us is alive and well.

Alan was quiet. But then, I have seen him that way before around big cats. Perhaps the small stuttering boy remembered the pacing Jaguar in the Bronx Zoo and his pledge to save him. Perhaps the indignity of the Tiger being gawked at by human beings in his forest home made Alan sad. I don't know, but I do know that as a scientist he appreciated the superb condition of the animal. The Tiger was about twelve years old, he said, and weighed five hundred pounds.

At day's end in India, with Alan Rabinowitz, whiskey, and a cigar

The Elephants finished their morning work, and after lunch we watched them bathe in the river, the mahouts scrubbing their thick gray hide with long tined brushes while they rolled on their sides and lifted their trunks, spraying water over themselves like a showerhead. A five-month-old baby took it upon himself to race out of the water and play with us on the shore, bumping his side against our legs and making circles around us like a big dog. The more we laughed, the more he engaged us body to body, until his mother began her ponderous wade toward the hilarity. The mahout intervened and pushed the little one back in the water, where he began to butt his patient mother instead.

Mahouts are born into their profession and a young Elephant is given to a young mahout to train, a relationship that endures until death. There are gentle mahouts and there are mean ones, just as with animal trainers anywhere. Elephants have been as indispensible to Asia in clearing the land as oxen were to North America. In Asia they are still used for logging and increasingly for tourists who want the experience of riding in a howdah and watching the animals perform. Thailand has few wild Elephants left in their forests, so in the northern city of Chiang Mai, a tourist mecca, the local Karen people have been illegally importing Elephants from Myanmar to the west, poaching their population of these remarkable animals. Life is hard for wild Elephants today; they are chased for capture or killed for their tusks. The working Elephant may have a better life; though enslaved, it is given food, a bath, and shelter.

The next morning I abandoned my birding buddies for early-morning Tiger spotting with Alan. As the sun was rising we saw a pair of huge Fish Owls, like our Great Horned Owl except for its exclusive partiality for frogs, fish, and other things aquatic. February is breeding season for owls in India, so they are seen in daylight more often.

Our jeep was not far into the forest when a mahout rode up on a small Elephant and led us to four Tiger cubs cuddled together like giant kittens in a jumble of lantana leaves. Alan said they were probably about eleven months old. One took off into the brush, then another and another while the mahout kept circling and circling around the snarling fourth until she too leapt away to safety.

As if that precious sighting weren't enough for a lifetime, an hour and a half later in an open field we saw four more cubs, thirteen months old, one gnawing on a Spotted Deer leg in the tall dry grass. They were lazy and full in the morning sun, staring at us with their honey eyes, the same color as their soft bed.

With abundant deer prey and no human predation, Kanha's Tiger population was increasing. Our group saw twelve of the cats in three days. They roamed the park and its buffer zones freely. Sometimes they wandered outside the zone and attacked villagers or their animals. As tribal villagers encroached more and more on the buffer zones, crossing through the park to get from one place to another, making contact with Tigers was inevitable, and it sometimes resulted in death for both.

"The number-one challenge for conservation in India is animal-human contact," said Belinda Wright, my dinner partner in Delhi one night. "One hundred and eighty-nine people were killed by leopards alone in 2003."

I was shocked. Millions of India's 1.2 billion people live next to forests and parks. Today the slow extirpation of predatory animals in contact with humans is a foregone conclusion despite the country's ancient respect for all living things.

In a new residential district of Mumbai, one of the largest cities in the world, twelve people, most of them children, were killed by Leopards in 2004, and twenty-two others were attacked. The Leopards' ancestral home was a sprawling nearby park, and one Leopard was on a ball field so often that it became the mascot of the ball team. As one mother lamented to the BBC: "It is really weird that in a city like Bombay [Mumbai] we have to live in fear of Leopards." Indeed! If even one person was killed by a Mountain Lion in New York's Central Park, that animal and all its cousins would be exterminated posthaste.

Belinda Wright founded India's Wildlife Protection Society. She is a photographer and filmmaker who has dedicated her life to saving her country's Tigers and other endangered animals. Her mother, Anne Wright, was on Indira Gandhi's original Tiger Task Force and helped create India's Wild Life Protection Act. India's wildlife laws are some of the strictest anywhere. A man who kills another man can get bail—but not a man who kills a Tiger.

India's population is poised to exceed China's in the next few decades, and land is being rapidly developed to accommodate the expansion, replacing wildlife habitat. The Leopard and the Tiger, squeezed into smaller and smaller habitats, are now prey themselves, falling to angry villagers and greedy poachers. It is getting harder and harder for Belinda's organization to safeguard the animals hunted in their own designated sanctuaries.

We saw the extent of the problem in the next Tiger reserve we visited. Bangalore, in the southern state of Karnataka in the Western Ghats, is a short plane ride from Mumbai. Ullas Karanth, a leading Tiger biologist with WCS, met us in the reserve, Nagarahole. It was fifteen degrees hotter than in Khana and the controlled burning along the forest roads added to the heat and smoky vistas. Still, graceful rosewood, teak, and sandalwood trees swayed in gentle breezes.

Ullas had visited my Putnam County home in New York when Alan brought him one day for a picnic lunch by our babbling brook. We walked in the lush deciduous forest hillside, where the Wood Thrush fluted and a Phoebe snagged flies over the water, and Ullas commented on the abundance of downed wood on the forest floor. Why wasn't it used by the people? We have more than enough, was my answer. "This would never happen in India," said Ullas. Later, looking around at White-tailed Deer picking up drop apples in the orchard, a Wood Turtle nipping watercress at the water's edge, a Water Snake sunning in the grass, and a Great Blue Heron darting fish in the lily pond, Ullas said, "All you need is monkeys in the trees." It had never crossed my mind that we have no monkeys in the United States.

Ullas's definitive study on threats to Tigers in the late 1990s revealed the extent of illegal hunting both within and outside all the Tiger reserves, mostly to satisfy China's insatiable appetite for Tiger bone for the medicinal market. In Nagarahole, Ullas was clearly worried about the poaching of all wildlife. Without enough prey, the Tiger would not thrive, and they were walking lucre for poachers. The Chinese market for Tiger parts was so lucrative that it was nigh impossible for the local people, including officials, not to be in on the take. Although naive villagers might kill a Tiger for a few hundred dollars, at the end of the line a Chinese merchant could get as much as $50,000 for the animal. The lustrous skin fetched a high price on the fur market, or as a wall hanging or rug; the bones were ground up and used as an elixir for just about anything. The penis bone is supposed to confer virility—although as Prince Philip is said to have quipped, "The Chinese don't need aphrodisiacs." The number of dead Tigers making their way to China did not satisfy the demand, so China began to farm them for market. In 1993, bowing to world pressure, China banned trade in Tiger bone and Rhino horn. Illegal poaching and trading was never stamped out, however, and, with the promotion of the State Forestry Administration, an estimated two hundred Tiger farms began to market body parts and Tiger wine. These farms hold between five and six thousand Tigers today, far more than those in the wild. If the government believed farmed Tigers would quell the need for the wild Tiger, it was gravely mistaken. The desire for a "superior" animal only increased illegal poaching of the wild cats for traditional medicine, wine, and pelts.

A government program called LIFT—Living Inspiration of Tribals—is responsible for the resettlement of tribal peoples out of forest reserves. On our 2004 visit, four hundred families had voluntarily relocated from Nagarahole, and eight hundred more were waiting. Each family received land, two bullocks, a house, and three years of government help to get started. One of the new settlements had neat adobe-like row homes with prosperous plots of garden vegetables and flowers, a baked-in-the-sun development like planned communities anywhere. I asked an older man, through translation, how he liked his new home. His cryptic reply was "My home is what it is." Then he added that he missed the breeze that came through his old thatched roof. He had gained the security of health care and better economics but lost the roots of his past. Unfortunately, men and wild animals cannot coexist in many wild places—we human beings keep taking too much.

Ullas had much to be proud of in Nagarahole. The wildlife had rebounded as people were relocated: Asian Elephants, herds of Gaur—wild cattle you do not want to confront on a forest trail if they decide to charge—families of Wild Boar rooting in the muddy banks, maroon-backed Indian Giant Squirrels in the trees, and hundreds of bird species.

One morning, as we waited patiently in the chill mist in our open jeep, the perfectly camouflaged face of a Leopard was illuminated low to the ground in sunlit vines. The cat cautiously emerged, deliberating whether he wanted to cross the road fifty feet from us, and then slunk back into the thick underbrush.

Leopards are more common than other cats in Asia, and far more secretive as well. They are magnificent creatures that can carry prey three times their weight up a tree to a large branch where it and they are safe from Tigers and Hyenas out to poach a meal. Leopards are the most adaptable of all the big felids, living in cold Russian climes as well as in the hottest deserts of Africa and Asia. Their adaptability has kept them alive, although the trade in Leopard fur and the fragmentation of their habitats continue to promote the decline of the species.

While no one wants or expects to live with a wild animal in a city—least of all the animal—out in the Indian countryside there are signs that education about wildlife is taking hold. Regular workshops have taught villagers to (1) listen for the warning cries of monkeys, deer, and dogs, (2) know the cats' routes, which are pretty regular, (3) clear the village periphery of grass, shrubs, and garbage, (4) go in groups when nature calls, and (5) use flashlights after dark to make the Leopard think you are looking at him. One villager claimed to know when a Leopard or Tiger was near because, he said, "He stinks terribly." This man even said that despite Leopard attacks he would prefer to live with the Leopard than without, because "tribals depend on the forest for their livelihood, and the health of the forest depends on the Leopard"—a sophisticated definition of the need for biodiversity.

Alan had no clear role on this Indian trip, other than as director of the Asia program, managing the work of Ullas and his colleagues. He gave a talk around the fire one night before dinner about animal-human contact and the bushmeat trade, but Ullas was our host and teacher. It was not Alan's usual place to take a backseat—nor was it George Schaller's.

George knew as much about Tigers as anyone in the world. He conducted the first scientific studies of the Bengal Tiger in India in the early 1980s. His book on the subject, *The Deer and the Tiger,* is considered the bible by many biologists. George could have spoken volumes about Tigers and the issues at hand, but with a grace that is uniquely his, he stood back and listened. I watched him in the lamplight, leaning against the wall, arms akimbo, as this new generation of Indian environmentalists took the lead with passion and integrity. His encouragement of younger scientists was legendary, from Jane Goodall to Dian Fossey to Alan Rabinowitz and now Ullas Karanth. His commitment to conservation was firm and clear:

If animals come into direct conflict with humans they need to be dealt with . . . The principal issue is to reduce conflict between man and animal. Regarding human rights and animal rights, you can do both, even in poor countries. The basic thing to remember is that we are wholly dependent on species, on the natural community for survival. When we destroy nature, we destroy ourselves. Everything we need, want, use and buy comes from nature. So we must protect it and all its species, but try to mitigate problems.

George Schaller has been one of the greatest influences in my life, not only regarding animals and conservation but also on how one behaves in the world. His tall, lean body treads lightly on the earth as if at any minute it might take him into the realm of the sky. He is soft-spoken, with a will of steel. In a room full of people he will speak of con-

George Schaller, looking spiffy for a night at the palace, Jaipur, India, 2004

servation as a process that is never done and can never be abandoned. He will talk of animals forthrightly, hiding the deep emotion that can just barely be glimpsed in his dark eyes. He is with human beings but his thoughts are always outside, on the vast plains of Mongolia or Tibet or Alaska, on the savannahs of Brazil, or in the bamboo forests of China, where he and his beloved wife, Kay, studied the Giant Panda in 1980. George handed out cards to hunters in those forests that said, "All beings tremble at punishment, to all life is dear. Comparing others to oneself, one should neither kill nor cause to kill."

We arrived the next evening at the palace of the maharaja of Jaipur. A flotilla of painted Elephants and Camels were lined up in the courtyard, with baubles, bells, and ribbons draping their bodies like so many Christmas ornaments. Had they ever known a life in the wild? It was doubtful. Their captivity spoke of an excessively opulent time, which I for one could not mourn. We were treated to a masterful dance in the moonlight by a eunuch—a skilled actor—another throwback to servitude. We dined in the flickering candlelight of a vast banquet hall, which had entertained throngs of dignitaries for 150 years.

I was sitting next to "Bubbles," the current maharaja, who in his seventies and having recently suffered a stroke was suffering my company as best he could, as I was his. Conversation was halting. The palace kept up pretenses like a tottering dowager empress whose painted lips and hair had seen better days. Bubbles needed the income we tourists provided. The inlaid mirrors and jewels, the

My friend Susan Sollins, me, and Bubbles,
the maharaja of Jaipur, at his palace, India, 2004

colorful cushions and bolsters that lined the floors, the dozens of
rooms flowing together under graceful arches, served only to remind
me of the time when thousands of Tigers were killed for sport, Elephants were broken as working animals, and the disparity between
the rich and poor was at its height. Today is a better time, for all the
battles waged.

I looked across the room at my fellow trustees, and my heroes
of conservation, George, Ullas, and Alan. How different it all is!
We come with cameras now instead of guns—most of us, anyway.
We believe in the rights of animals and of human beings. I watched
Alan a few tables away. He was sitting back in his chair, quietly, a sad
look in his eyes. Perhaps he longed for home and the sweet reward
at day's end of Alex and Alana in his life, or maybe the cancer was
wearing him down. I suspected otherwise. My hunch was that he was
trying to figure out how Tigers were going to survive in their island

reserves within the multitudes of villages pressing on them. What kind of genetic health could they sustain over the years? Surely they would go extinct like the pygmy Taron people of Myanmar without access to new genetic strains. How do you keep Tigers forever? The answers would bring him full circle.

8

Brazil

The Pantanal is the Serengeti of South America. Nowhere else can you see wildlife so openly and in such abundance and diversity. While the great plains of eastern Africa are savannah, extending in endless profusion across Tanzania and Kenya, the Pantanal is the world's greatest wetland, comprising an ancient basin that fills and empties with the seasons in Brazil, Bolivia, and Peru. There is immense diversity of life throughout Amazonia, but in the Pantanal, with its long low vistas, it is possible to really see the animals.

As a girl growing up in Boston I had a Eurocentric view—a legacy of the Founding Fathers, who looked to the great cities across the Atlantic for heritage, culture, and beauty. I spent my junior year abroad at the venerable four-hundred-year-old University of Edinburgh and tromped through the Scottish Highlands and outer islands in my time off. There is nothing quite like the sweet smell of crushed heather underfoot, and to this day I am inclined to head north for new discoveries, and a glimpse of Curlews with their long curved bills plucking berries from the heath.

No one ever talked to me about the huge expanse of the Southern Hemisphere, although I stared at it on the globe in my father's study. No one I knew ever went there. They visited Havana for gambling and big bands, and the Caribbean once in a while, or Africa for a safari if they were really flush. But Mexico began calling to my friends and me like a colorful cheap bouquet, and it was just a border hop away. I made multiple trips there in my twenties, camping in the purple deserts of the north, pushing my toes through sandy beaches in the west, or riding horses in the mountains of the south. The people were kind and the food was delicious—once the requisite bout of Montezuma's revenge was endured and over with.

I finally ventured farther south to Belize and Guatemala, encountering exotic birds, Mayan temples, and reefs along the coast with an underwater life so miraculous it seemed like a dream. In Guatemala with Victor Perera, we encountered a war-torn people, sad and deeply weary from burying so many loved ones. Guerrillas hid in the forest while government soldiers ripped the film out of my camera before I had a chance to document a village rout. A woman wailed in the early morning light as our boat docked on Lake Atitlán. She had witnessed the horror of her husband being gunned down just hours before. In a Guatemalan refugee camp across the border in Belize, children ran to us through the mud, their bellies swollen with pellagra and their eyes clamped shut with infected fly bites. They had been there for years with no hope of getting out. The civil war lasted from 1960 to 1996 and was essentially genocide of the Mayan people by a repressive military regime. We spent a day with Rigoberta Menchú, a young K'iche' woman from the Guatemalan highlands whose parents and brother had been tortured and killed. She told her story in *I, Rigoberta Menchú*, which was published in 1982 when she was just twenty-three years old. In 1992 Rigoberta won the Nobel Peace Prize and continues to this day to bring justice to her fellow Guatemalans and hope to indigenous people the world over.

I kept looking south, and when asked to cruise the Amazon on a theater trip, from Rio to Manaus, two first-class tickets in exchange for three performances, I jumped at the chance. Brazil was everything I had imagined, read about, or dreamed of. The river was overwhelming in size, like crossing an ocean, with no visible shoreline for hundreds of miles. The occasional Green Iguana rode the waves downstream on earthy islands ripped from the land, and the murky reddish waters carried more volume than all the ice locked up in

Greenland. When at last I did venture into the jungle interior, it beckoned like a magnet, promising a wealth of exotic species, which always delivered. I understood what early explorers must have felt and why they needed to continue on up the Rio Negro or down the Paraguay, butting finally into the impenetrable snowcapped Andes, the source of it all.

Once I was bitten by the lure of South America I kept returning: Brazil, Ecuador, Colombia, Peru . . . it was not possible to put my arms around the vastness of the experience. I was barely dipping a toe in. When Alan invited me to go to the Pantanal in 2010 to see Jaguars I was thrilled.

Cats are notoriously hard to see in the wild. They are masters of stealth, stalking and then pouncing on unaware prey. They usually work alone in high grass or forest cover. Lions and Cheetahs, which kill communally, are the exception. Alan would rarely see his study animal except in capture for radio-collaring or on critter cams. He was shocked once to learn that the Jaguar he was following in Belize had circled round and was actually following *him*.

I never saw a Jaguar when I was with Alan in Belize, nor on subsequent visits. We saw Tigers in India, routed out by canny Elephants, and I spotted a Golden Cat in the high forest of Bhutan. In Africa, prides of Lions took down prey on the open plain and Cheetahs raced after Zebras for all to see. On safari it is expected that the wildlife will cooperate.

One of my best cat sightings, however, was totally unexpected and close to home. I was birding a mile from my house under some power lines bordering a New York state park. It was a hot August day and I lay on a high granite boulder to rest. When I sat up twenty minutes later, there under the power lines, rolling in the dusty path about two hundred yards away, was a Mountain Lion, or Cougar, its long tail dotted with bits of black, marking it as a juvenile. Still, it seemed full grown, and I watched in fascination as it rolled and rolled, covering its body with a shower of dust the color of its fur. Finally it rose, shook itself of the cloud, and wandered slowly into the woods, never sensing my presence.

Alan lived nearby, and he and I were soon out tracking the cat. We found scat with deer hair in the feces and Alan saw what looked like a scrape on a tree, the claw striations that marked territory. The area was perfect habitat, many crevices in granite outcroppings left by the last glaciation, and abundant White-tailed Deer, wild Turkey, and Cottontail Rabbit. The U.S. Fish and Wildlife Service has not acknowledged the recolonization of New York and New England by the Mountain Lion, despite ten thousand reported sightings since the 1960s. The cat had been extirpated in the 1930s, they told me; the one I saw was probably the release of an ill-advised pet owner. Whatever it was, it was real and marvelous to behold.

In September 2010, I flew with Panthera's chairman, Tom Kaplan, in a chartered jet over the Amazon. The flight from Teterboro Airport in New Jersey to the city of Cuiabá in Brazil took six and a half hours, with no time change. The journey was over green, green, and more green, the greatest expanse of it in the world. An occasional flare of smoke or fire attested to the slash-and-burn clearing of the land common to tribal people, but for the most part the verdant blanket was uninterrupted, except for the cut of the mighty river west to east.

Tom Kaplan fell in love with gold as a teenager, after he decided the stock market was a kind of giant Ponzi scheme with little tangible to bank on. He traded wisely and ended up distributing his wealth to favorite causes: health, the humanities, and the survival of endangered species. In 2006 he founded Panthera, dedicated to saving the wild cats of the world. He had followed Alan's work with WCS and asked him to lead Panthera as its CEO. Together they recruited top field biologists, including George Schaller, and began producing the best science and strategies to save the big cats. They asked me to join the advisory Conservation Council, along with my fellow thespian and conservationist Glenn Close, and I was now going to Brazil to see the work they were doing.

As we put down on the small grass landing strip, old friends Howard Quigley and the photographer Steve Winter greeted us, along with Alan. Bob Simon and a crew from *60 Minutes* were there as well, shooting a segment on Alan, Tom, and the Jaguars. At lunch I learned they had seen Jaguars only at night and were getting a bit desperate for day footage. Steve told Bob he once spent three months waiting for a Jaguar photo with no luck. It was too hot to go out between 1 and 4 p.m., so we all retreated to our small brick cabins and waited for the 114-degree temperature to subside.

I crossed the dusty central plaza of the old fishing camp, the Porto Jofre lodge, skirting the horses that freely wandered the grounds, as did Crested Caracara and six-foot-tall Jabiru Storks. A gigantic fig tree dominated the plaza and housed many chatting Monk Parakeets, Black Vultures, Horneros, Kiskadees, and, best of all, six Hyacinth Macaws, which jumped lazily from branch to branch gobbling figs. The largest of the macaws, the endangered Hyacinth looks like a cartoon cutout of a bird, with a prominent yellow ring circling black inquisitive eyes. The hyacinth of their name is more electric blue than lavender, and the yellow plumage on their face heightens the effect. They roost and nest in large tree holes and pop their heads out occasionally like a jack-in-the-box. They are magnificent birds, threatened by the pet trade and for their feathers. Here they were safe.

In the late afternoon the CBS crew climbed into one of the motorized pontoons and the rest of us followed, several to a boat, speeding up the Paraguay, the wind on our faces a welcome relief from the oppressive heat. I was with the Brazilian biologist Ricardo Boulhosa, a specialist on Jaguars who also knew his birds. Our driver was a handsome young man named Mateus, an uncanny spotter of wildlife. We veered from the others toward a small tributary, passing Amazon Kingfishers spearing small fish, Cocoi and Tiger Herons standing in the shallows, and Large-billed Terns and Black Skimmers sitting on the sand. A family of rare Giant River Otters cavorted in the water while a lone male downstream chased a caiman with a large striped catfish in its jaws. The Otter actually tried to grab the catfish out of the caiman's mouth when suddenly another caiman bellied off the sand into the water and chased the Otter away. A reptile protecting another reptile! What, I wondered, did the second one get in return for this favor? A good chunk of fish? I am always astonished at interplay within and between species.

A family of Black Howler Monkeys, among the largest monkeys in the New World, was resting in a big tree. The littlest juvenile was hanging off a branch swinging at the tip when two large vultures flew in on the branch above him. That's when the biggest male Howler, the size of a three-year-old boy, who was munching on leaves twelve feet away, lifted himself and unhurriedly crossed the branch hand over fist to swat at the vultures, just as any dad might do. The huge black birds rose in the air and took off as Dad retreated to his comfortable perch, a protective eye still on his little one.

Mateus cut the motor and signaled for quiet. A tangle of vines and brush fifteen feet above the earthy loam of the riverbank was lit by the low sun, and in a cleared spot beneath the shade of a large tree lay a Jaguar grooming her legs and chest. A glint of light pierced her eyes as she looked up briefly and registered us not thirty feet away. She seemed calm, not fearful or even curious, and returned to her grooming. Maybe she knew she was safe on land with us on the water. Or maybe it was just too hot to stir herself. The upwelling of the river breeze had brought her to the spot and there were hours to go before she began her hunt.

The Jaguar seems perfect in its beauty. Her body was taut and muscular, the coat hugging her sinews like the sleek costume of a trapeze artist, or a wrestler. Each black rosette of her fur was an exotic flower dotting a rare field of beige. Jaguars have large heads with the most powerful jaws in the world. The name "Jaguar" comes from an Indian word, *yaguer*, which means "killing its prey in a single bound." They leap on the back of peccary, deer, or Capybara and crush the skull or spine with their canines. The males can grow to seven feet long and weigh two hundred pounds. They are not aggressive toward humans or other predators, preferring flight to fight. I never tire of watching them, although it has been mostly in zoos, where they are not content. They are solitary creatures, for the most part; confinement does not suit them.

I was finally seeing one in the wild, truly in the wild of Brazil's

Alan Rabinowitz and his wife, Salisa, on top of the world, Nepal, 1990

Tiger in Kanha, India, 2004

Hyacinth Macaws at their roost hole in the Pantanal, Brazil

The first Jaguar I ever saw in the wild, Brazil's Pantanal, 2010

In the mountains of Peru on tree-planting day, 2006

The Marvellous Spatuletail Hummingbird in Peru

A clansman in Papua New Guinea sports many kinds of Bird-of-Paradise feathers, including the longest tail feathers in the world, the white ones, from the Ribbon-tailed Astrapia, 2010.

He looks fierce, but this Marine Iguana from Galápagos is all bluff.

The Uakari,
a primate from
the flooded
forests of
Mamirual, Brazil

My grandsons: Mac finds a small chameleon in Madagascar,
and Finn snaps a shot, 2015.

The Fossa is one of the few carnivores in Madagascar—
eating lemurs mostly, 2015.

My husband, Ed Sherin, and me in heaven in Bhutan, 2012

The bird I most wanted to see in the world,
the rare Monal Pheasant, Bhutan, 2012

My first Jaguar in the wild, Brazil's Pantanal, 2010

Pantanal and not in a zoo or reserve. We observed her for a long time. I looked up from my camera lens and found her staring at me, the same look countless native women have given me when I dare to snap their picture on a foreign street. It was discomfiting, and I suggested we leave her to the peace of day's last light.

We were late coming to dinner. Alan rose when he saw me enter, seeing the excitement on my face. I embraced him in gratitude for the incredible sight of the Jaguar, almost thirty years after we had first tracked them together in the jungles of Belize. I described her and the abundance of wildlife on the remarkable river. It turned out that Ricardo and I were the only ones who had seen a Jaguar in daylight. The *60 Minutes* fellows had one day to go. They scrambled to get out on the river.

The Jaguar of Central and South America is a single species in its entire range, unlike the Tiger of Asia, which has many subspecies, three gone extinct in the past eighty years. Only the Cougar has a greater range throughout both Americas. The Jaguar was exterminated in the United States by the 1930s; Alan believes that the few still found in Arizona are border crossers from Mexico seeking a mate, not a viable population yet. A male Jaguar travels extensively in search of new territory and a mate. Panthera's studies found that they will go five hundred miles and more in their search, which ensures genetic diversity.

Although the Jaguar is better off than the other large cats in Asia and Africa it has still lost 60 percent of its range to development, mostly the huge cattle ranches throughout the Pantanal called fazendas. The ubiquitous white Brahma Cattle grazing these ranchlands surround the Porto Jofre lodge. In fact, Tom Kaplan bought the fazenda across the river from the lodge to keep Jaguars from being shot for killing cattle, an ongoing problem if prey is scarce. Taking down a full-grown steer is not easy for a Jaguar—they would prefer peccary, Capybara, or even caiman—but if a calf or a diseased cow is available they will kill, and then be killed in return by the gauchos. Tom taught his ranch managers that the tourist trade was more lucrative with Jaguars alive. The idea was to replicate this policy with all fazendas in Brazil, and to reimburse the owners for cattle the Jaguars kill until the income from tourism balances out. It is beginning to work. Ranchers are opening their fazendas to tourists, who have the best chance anywhere of seeing the great cats.

The night air cooled to the nineties and our four boats motored up the river in the dark. The glowing red eyes of caiman broke the surface of the shallows as we buzzed past. A Tapir slipped into the river for an evening swim, unfazed by our presence. Something nipped at her feet and she splashed to safety on the other shore, turning with a last querulous look. All the mammals seemed larger here: the Capybara, already the biggest rodent in the world, seemed giant in comparison to those I had seen on the Rio Negro or in Belize, like an overgrown Muskrat crossed with a huge rabbit. The Tapir, the River Otter, the Jaguar, all thrive in this world of plenty.

Mateus spotted more Jaguars on our excursions than anyone else, and it was getting embarrassing, especially as some people had seen none. This has happened to me before—in fact, many times. I have

The largest rodents in the world, Capybaras,
take a break on a road in the Pantanal, Brazil, 2010.

uncanny luck seeing rare birds, reptiles, amphibians, and mammals wherever I go, so much so that others have asked to join up with me, as if I were a good-luck charm.

By the fourth day I had seen five Jaguars, while the TV crew and some Panthera board members had to depart without a single day-time glimpse. They did have a remarkable night encounter, however. A Jaguar with a cub entered the water and circled around a caiman like a cowboy rounding up strays. She then pushed the caiman toward shore, pounced on its back, crushing its skull, then flipped it over in the sand and ripped at its throat. With the cub trailing behind she pulled the carcass up the sandy bank. Alan was very excited by this sighting, as he had never seen such behavior in a Jaguar before. The cats are great swimmers, but she was clearly teaching her cub how to drive the reptile to shore.

The fazendas are so extensive—sometimes they encompass a thousand square miles—that Jaguars can easily traverse them without being detected. Alan was concerned that with encroaching development along the cat's full range, especially in Mexico, Central America, Colombia, and Venezuela, the animal would be hindered from free travel and ultimately compromised in genetic diversity. He had seen this happening with Bengal Tigers in India. Jaguars were doing well—there were perhaps tens of thousands of them—but the time to protect their future was now, before dispersal routes were blocked. So Panthera created the Jaguar Corridor Initiative.

In the 1980s when I was at Conservation Committee meetings at the Bronx Zoo, a biologist on staff named Chuck Carr was involved with a Central American plan to create a corridor for wildlife called Paseo Panthera. Alan and I admired the concept and spent hours talking with Chuck about it. It was a far-thinking idea, which was ultimately diluted in bureaucracy but had real value Alan never forgot. The Jaguar Corridor was a grander scheme, and one that would require decades to complete.

Panthera hired a young woman named Kathy Zeller to map dispersal areas of existing Jaguars from Arizona to Argentina, and subsequently layer the maps with viable paths and areas of concern. Howard Quigley's research uncovered surprising facts. Jaguars were moving through relatively small and densely populated human areas to reach their destinations, and not only at night. The Jaguar's own predilection for nonaggression toward human beings is probably what has made it so successful as a survivor. It stealthily crosses human habitats without detection, not stopping to take on dogs or other possible aggressors. George Amato had moved his genetics and genome labs to New York's Museum of Natural History, founding the greatest repository on earth of animal DNA—more than 100,000 species, from dinosaurs to the great cats. Alan's wife, Salisa, helped him map the Jaguars genetically, exposing their family structure and health along the corridors of Central and South America.

After the science was complete, Alan began meeting with politicians, ranchers, and local communities in the countries where Jaguars range. By 2015, Panthera was working with people in fourteen of those eighteen countries. The work was wide-ranging and included landowner assistance and training, mitigating human-Jaguar conflicts, and opening possible new corridors, such as connecting the western population of cats with the eastern over the Andes in Ecuador, through memorandums of understanding with government leaders. Panama became crucial in this regard, as the slim isthmus is the only corridor connecting Central with South America. Local response to the Jaguar was surprisingly positive throughout its range. There is a deep cultural connection, after all, between the native people and the Jaguar, a demigod for ancient cultures such as the Olmec, the Maya, and even the Aztec. The cat was revered as the ruler of the night and the underworld. Despite the decline of these beliefs, the respect for the Jaguar persists in new generations, making their protection an homage to their ancestors. The Jaguar Corridor is the first large undertaking of its kind and has become a model for migrating species worldwide.

My admiration for wildlife photographers is immense. They brave weather conditions that no sane person would endure, often alone and for months on end, to get a shot of an elusive or rare animal. Steve Winter is such a photographer. I first became aware of his work in *National Geographic,* and when he shot Alan for an article I got to know him. He and his life partner, Sharon Guynup, a journal-

The wildlife photographer Steve Winter
checks his camera traps in Brazil's Pantanal, 2010.

ist, are staunch protectors of endangered species. Their book *Tigers Forever* is one of the best on the subject, with Steve's iconic pictures and their cowriting.

Steve and I went out on the fazenda one searing afternoon to inspect his camera traps. He was hoping for cat photos, and what he had was 161 pictures of cattle crossing the frame. He chuckled at his bad luck and reset his cameras; as we ambled back toward the river, we were rewarded with the sight of a tall graceful Rhea, South America's Ostrich, with nine fuzzy chicks racing after her.

At the water's edge was a caged barge crowded with cattle to be shipped downriver that night. The sun was beating down on the animals and several had sunk to their knees in the heat. I didn't think many of them would survive and alerted Tom to the impending deaths. Indeed, within a few hours, by the time the ranch hands reached them, nineteen had perished. And they worried about the occasional Jaguar kill?

Our days in this Eden of Brazil were coming to an end. Ricardo and I were joined by another Conservation Council member, Jonathan Powell, who had been Tony Blair's closest adviser when he was Britain's prime minister. Jonathan was keen on birding and had also heard that we had seen eight Jaguars. The three of us set out at dawn to catch the early birds, then Ricardo drove us to out-of-the-way places such as his old study site deep in the forest. It was a ghost laboratory, like something out of a horror movie. Specimen jars sat on rickety old shelves covered with dust: baby reptiles caught forever in formaldehyde, snakes, frogs, eggs, and insects all in a row as jungle vines crept over them. A table was littered with mammal skulls, eye sockets eerily caught in the midday light. Vines crept through cracks in the wooden roof and would soon envelope all traces of the lab. Ricardo had left it ten years earlier to go south for a new position. We moved on to visit an old rancher and his wife in their neat small house in the middle of the vast dry landscape. They were friends of Ricardo's and were delighted to welcome him again; they served us shots of guarana, a coffee-like powder mixed with water and sugar, which gauchos drink all day for the boost it gives.

A twelve-foot-long Anaconda inched across the road as we drove away, its girth that of an NFL linebacker's arm. Capuchin Monkeys played in the brush while Jabiru Storks nested in the highest tree, three chicks peeking over the edge. Every morning or evening we would see another Jaguar, always on the riverbanks taking the cool breeze. This was the richest environment for wildlife I had ever encountered outside Africa.

On our last evening Tom asked me to go fishing with him. My luck held even though I am not much of a fisher. I reeled in a ten-pound sleek silver catfish, one of many species of catfish in the Paraguay. When Tom landed a smaller gold-speckled catfish, I told him it meant that the price of silver was going to rise. Tom had initiated many probing conversations about life, philosophy, and conservation during our time together. He is a born historian and a committed environmentalist, who generously made it possible for us to come to the Pantanal and witness the work of Panthera.

Alan's star kept rising. The *60 Minutes* segment was very popular with viewers. *National Geographic* did an hour-long movie about

Panthera's leaders, left to right: Luke Hunter; Alan Rabinowitz; Rafael Hoogesteijn, a Venezuelan Jaguar biologist and cattle rancher; Tom Kaplan; and Howard Quigley

him, as did the BBC and others. Readers love his books because he couples his passion with exciting personal adventure stories about wildlife.

The little boy from Queens still stutters a bit, which serves only to make him a hero to fellow stutterers and organizations like the Stuttering Association for the Young, which he wholeheartedly supports. His leukemia may be slowing him down physically; mentally he is accomplishing more than ever. "But at my back I always hear Time's winged chariot hurrying near," wrote Andrew Marvell. Time is short—both for the animals and for Alan.

His legacy is written in the policies of the countries where he has worked: Belize, Thailand, Myanmar, and those of Central and South America in particular, which now all have protected areas and laws in place for the great cats. In Asia and Africa the cats will survive in large parks if enough enforcement is in place to deter poachers. In the New World, Jaguars and Cougars will roam free if the corridors survive human development and are kept open, and as long as the political and social will is there to allow them their freedom.

Alan's contribution to wildlife conservation is immeasurable. He is a hero for all time. If the great cats of the world survive the onslaughts of this century it will be because of my friend the Tiger Man, the little boy who grew up to keep a promise he made to a Jaguar more than fifty years ago.

PART 2

————

WILDLIFE WOMAN

9

Birds

We lived on Pill Hill, where almost all fathers were doctors. Hawthorn Road was a short street with just eleven houses. My dad finally came home from the war after five long years with the Fifth Army Hospital in England, and he and my mother bought the old clapboard house in Brookline, Massachusetts, for $10,000. Those postwar years were special. The country was in a celebratory mood, and for families reuniting there was joy in the air. We kids spilled out of our homes to play Kick the Can before supper or to race down Cumberland Avenue to look for turtles or minnows in the Muddy River shallows.

There was a vacant lot we called "the woods" catty-corner from us where I climbed trees, examined bugs, and first became interested in the natural world. The intense fear I had of spiders dissipated when my mother took a huge spider from under the back porch eaves and let it crawl up her arm and over her

My life as an adventurer began at three years old.

head. She didn't flinch. I thought she was the bravest person in the world. In our little plot bordering Boston, she taught me what she had learned of nature as a girl in Nova Scotia, and she began to name the birds. There were starlings squeaking noisily in the locust tree, a Mockingbird perched on the streetlamp, and a Robin who woke me each morning with his sweet song.

At the end of the street was a clay tennis court where our neighborhood gang hit the ball around when the weather was good. There was always a flock of House Sparrows there, snug against the chicken wire eating something. At ten years old I became intensely curious about flight. I jumped off rocks and bluffs trying to fly, and made balsa-wood wings to aid in the effort. My favorite story was the myth of Icarus, who flew on wings of feather and wax too near the sun and plummeted to the ground. I loved the illustration of him in the book of Greek myths Dad read to us before bed. Icarus was like a great eagle soaring toward the sun, his wing tips bowed against the air. It was the ultimate picture of freedom. I guess my brother Tom loved it too because in his twenties he joined the Marines, served in Vietnam, and became a pilot for life.

The little sparrows gave me an idea. I took my clunky balloon-tire bike up the embankment on the far side of the tennis court and then pedaled like mad across the court in ambush. It worked! One sparrow couldn't get up in time and was caught in the chicken wire. I held the creature in the palm of my hand and felt its trembling heart. The bird was so light. I looked into its eyes, its beak yammering in protest, then turned it over on its back and extended one wing, examining the structure. I turned it over again, opened my palm, and watched as it magically lifted into the sky. Perfection.

Puberty hit with a vengeance. Icarus was replaced by the goddess Aphrodite, and sparrows by boys. I discovered acting. The woods were replaced by theater and an equally enchanting world of make-believe. I did not get back to birds for twenty years.

My career was well established when Ed and I bought our house in Putnam County, just sixty miles north of New York City, a place where our four boys from our first marriages could play and grow. A gracious lawn rolled down a slope past a lily pond over a brook to a pasture and an old orchard of twenty trees. Behind us was a forest bordering on state conservation land. Wildlife was everywhere.

A herd of White-tailed Deer wandered the orchard eating drop apples. Occasionally a River Otter slipped into the brook's pool, created by the dam of an old sawmill. He splashed for Brook Trout and Water Snakes. Opossums cornered by our curious dog played dead in the woodpile. Rabbits made smooth bowls in the earth beneath cat briar and birthed a dozen bunnies. Wood Turtles with bright orange legs bred in the wetlands and baby turtles could be found nibbling on pasture flowers in May.

One spring day a family of Barn Swallows came wheeling up the lawn from the South, circling excitedly round and round like kids tumbling out of a station wagon at their summer home. They nestled under the back deck and busily built mud cups for eggs, as they had for all the years the previous owners occupied the house.

A pair of Eastern Phoebes, sweet black-topped flycatchers, arrived on March 25, as they did every year afterward, give or take a day or two. They staked out a little shelf under the roof overhang and built a nest of moss and grass and mud. Their soft *fee-bee* call was comforting. They darted out from a branch above the stream or the pool and grabbed little white moths or flies for their chicks. It took hundreds to fill the four gaping mouths every day.

By the second year of living in our "Pumpkin House"—so named for the big window in front and the two above on either side, which made the house look like a jack-o'-lantern lit up at night—it became clear to me that we shared our place with hundreds of species—or thousands, if you included insects. And all these creatures predated us in lineage of the land—maybe for millennia. We were the interlopers, but we would be careful not to interrupt the conditions that allowed them to thrive.

They became friends and, perhaps sensing my joy in them, moved closer: the House Wren, the Robin, the Phoebe, and Barn Swallows practically lived in the house with us, and in time the little Brown Bats *did* live in the house with us, under the roof, while the slender Ring-necked Snakes birthed their young in the foundation.

One day I opened an unzipped suitcase in a closet to find an exquisite Flying Squirrel protecting four pink babies on ragged tufts of mattress stuffing. As much as I admired her beauty I could not see us sharing the cathedral ceiling space with five Flying Squirrels, and so began the comical attempt to evict her. I put on thick ski gloves and innocently went to pick her up, whereupon she dashed out on the second-floor railing and leapt to the fireplace, climbing the stones all the way to the ceiling.

I probed the distance between us with a long tree clipper, which induced her to glide gracefully through the air to the floor twenty-five feet below. Then she ran up the spiral staircase toward her babies as I raced to close our bedroom door. This scenario repeated itself for the better part of two hours.

She was the most beautiful rodent I ever saw, with big brown nocturnal eyes, a gray-brown back, and a cream-colored belly. When she opened her paws to "fly," the expanse of soft furry skin from hind leg to foreleg was pulled taut and allowed her to sail with ease through the air like a paper plane, angling this way and that with her wrists before landing gently on the oriental rug. I was entranced. I caught her a few times but she struggled free after vigorously biting my thick gloves. At last I caught her in a butterfly net and shut her up in a cardboard box.

I moved her babies down to the basement, cradled in a towel, and put a plate of nuts and fruit next to them on top of the old foundation. Then I released their weary mother nearby. The stones had so many cracks between them I suspected she would find her way out. She did. When I returned a few hours later her tiny pink babies were gone and she had eaten every bit of fruit and all the nuts. I half expected a thank-you note like Santa leaves.

The birds continued to enthrall me the most as the seasons went by in our country home. The Great Blue Heron, sensing sanctuary, barely budged when we ran down to the lily pond hoping to keep just one Goldfish alive after the rest had been devoured. We saw the looping display of the Woodcock courting in the orchard and the Louisiana Waterthrush, which is not a thrush at all but a warbler, teetering at the edge of the brook, its sweet song burbling against the water's lap. The fuchsia-red blaze against the white of the Rose-breasted Grosbeak looked like a crucifix on an Easter chasuble.

While I came to know the birds at my feeder best, there was one stationed fifty feet into the woods that kept his distance. This was a bird whose song transported me. I waited for it in the morning before I rose and in the evening at dusk. It is the bird I love the most, and the song gives me chills. It seems of another realm. We were lucky to have a Wood Thrush call our place home. In the thirty-four years we lived there we had several generations of Wood Thrush songsters, and I can say emphatically that there are singers and then there are *singers*.

At some point the one I called "the grand old man" began to sing. He was with us for five years. His song was to the previous Wood Thrush as Pavarotti's is to Garth Brooks's: both are good singers, but the former takes your breath away. I spent enough time listening to the grand old man to begin to believe that he knew he was remarkable. I had the feeling that perhaps the whole forest stopped to listen. He began to improvise, and the song became more than the standard "*e-olay.*" It became intricate in its dynamic, one phrase drawn out and full-throated and the next rippling down softly as if spilling from a bowl. He had a deep alto resonance. He was raising the bar in the Wood Thrush world.

One evening an astonishing event occurred. It was one of those perfect late July evenings, golden light and about seventy-five degrees. We were sitting on the back deck when the grand old man tuned up. He sang gloriously for about five minutes and then suddenly was joined by not one but two others who must have been sitting about ten or twenty feet on either side of him. He would begin and then the second one would come in softly and then the third, almost as if singing a round, but then they deviated. They would truncate phrases, pull a note out of a high register, take off from one another, undercut with a low note, one doing a percussive rat-a-tat chip-chip as counterpoint and then they would end with a flourish before starting again. This went on for about twenty minutes—a Wood Thrush jam session. Jazz in birdland—worthy of Charlie Parker or Miles Davis. Were the other two learning from the master? Were they juveniles learning from Dad? If so, they were learning without any prior lessons that I had heard. I never knew the answer but in the two weeks that followed we were treated to three more concerts and then nothing after August 3, 1999, although the grand old man was with us for a few more years.

Succeeding thrushes never equaled his song and in fact some were almost screechy. The grand old man was my "spark bird," the one who turned my love for birds into a passion. I never look or listen to a bird today without seeing or hearing its individuality—especially if I am lucky enough to live with them nearby.

Here is a bit of a poem by the nineteenth century's Gerard Manley Hopkins, on the English Song Thrush:

> *Nothing is so beautiful as Spring—*
> *When weeds, in wheels, shoot long and lovely and lush;*
> *Thrush's eggs look little low heavens, and thrush*
> *Through the echoing timber does so rinse and wring*
> *The ear, it strikes like lightnings to hear him sing . . .*

> *What is all this juice and all this joy?*

Indeed! What is all this juice and joy? Hearing the grand old man sing was to believe he did it for the joy of it, especially in August after the breeding season. The English Thrush cannot compare with our Wood Thrush of the Western Hemisphere, but thrushes the world over pour out their joyous morning songs to the delight of human beings everywhere.

The Audubon Society has more than 460 chapters all over the United States. I joined the Putnam Highlands chapter hungry to learn about birds by going out with folks who knew them. An older

man named Tom Morgan took
me along on May mornings to
spot warblers migrating north.
Tom was bent over from osteo-
porosis, making it hard for
him to look up, so he knew
the songs of all the birds—
hundreds of them. It was a
wonder to me how he distin-
guished between a Magnolia
Warbler, a Yellow Warbler, or
a Redstart. We would sit on a
granite outcropping in the hilly
terrain of Fahnestock State
Park and Tom would tell me
about the birds: "Hear that? A
Canada Warbler. And there's
a Fox Sparrow, and a Scarlet
Tanager." I was hopeless at the
art for many years.

*Tom Morgan, who taught me how to
listen to the birds, Putnam County,
New York, 1975*

When I was asked to join
the Gone Birding team for the
World Series of Birding in the 1980s I thought they were kidding.
However, as the on-camera host of the Gone Birding! video game,
I had developed a reputation as an actress who might help promote
the cause of protecting birds.

I found myself in a car with four ace birders at midnight in a
swamp in New Jersey next to I-95, where refineries belched smoke
and flame. We had twenty-four hours to see or hear as many species
as possible throughout the state. This little patch of brackish water,
left over from the ancient coastal marsh that had been sucked up
by mankind's need for energy, happened to house the Black Rail, a
tiny secretive denizen whose metallic call rang out loud and clear in
hopes of attracting a mate. It was our first bird of the marathon, a
little gem in the middle of hell.

We zoomed down the highway to our next spot, probably for owls, and so it went through the night until dawn brought us more birds than we could handle and the thrill of rare ones we never expected. We visited wetlands and forests and shorelines. The count grew as the day wore on, and our weariness increased.

I was in awe of my colleagues, especially Julie Zickefoose, whose watercolors and stories of rehabilitating injured and orphaned birds have become legendary. She was in her twenties back in the 1980s, and she had an ear to die for. She would roll down the window, cup her hand to her head, and call out: "Carolina Wren, Great Crested Flycatcher, Swainson's Thrush, Fish Crow . . ." as we sped by. This crazy chase left no time for lunch, or even relieving oneself except on the fly, pun intended. We staggered into Cape May headquarters late Saturday night with our tabulation. It was a respectable count, probably 170 or so species, but far from the best because I was almost dead weight for the team. There was a lifetime of learning ahead. I began to listen, really listen.

We may never know what Pterosaurs sounded like, but paleontologists have a pretty good idea what they looked like from examining their fossilized bones. They were reptiles that flew in the Jurassic and Cretaceous periods, 200 to 65 million years ago. To put it another way, they were on earth for around 135 *million* years, 134.8 million years longer than modern human beings. They were far more diverse than the line of flying dinosaurs that evolved into our birds.

There were hundreds of Pterosaur genera, some with a wingspan of only ten inches and some with a wingspan of thirty-six feet, the first fossil of which was discovered in Big Bend National Park in Texas a few decades ago. They came with elaborate tails, crests, beaks, and even plated forearms on which they walked. If you have seen the animated movie *How to Train Your Dragon 2* you get an idea of what the skies might have looked like when Pterosaurs filled them. Flying reptiles everywhere.

The Cretaceous was a warm period with a mean temperature

The World Series of Birding team, left to right: Me, Ollie Komar, Peter Alden, Alf Wilson, and Julie Zickefoose, New Jersey, 1980s

above 60 degrees, a carbon count of 1,700 parts per million for much of the time (today we are at 400 ppm), and, in the late years, a higher oxygen level than we have now. It was a wet world, a fecund breeding ground for reptiles, shellfish, and Pterosaurs, which ended when a six-mile-wide asteroid or comet collided with Earth in what is now Chicxulub, Mexico, killing most creatures, all the nonavian dinosaurs, and all the Pterosaurs. At least, that is the hypothesis. No event 65 million years ago can be proved to have taken place, but rocks the world over speak of something catastrophic happening after impact. The crater it made is 110 miles wide and the collision was equivalent to a billion atom bombs.

While it lasted, the flying experiments called Pterosaurs were astonishing, as were the variety of dinosaurs. But the Cretaceous period was particularly fertile in many ways. Flowering plants began to develop. The ginkgo trees that line New York City streets today were planted because they can withstand carbon overload and pollution, as they must have 100 million years ago. Ants, butterflies, and bees evolved, making use of the flowers and trees. They survived the big extinction event 65 million years ago, as did marine reptiles, snails, clams, amphibians, lizards, and snakes. Omnivores, insectivores, and carrion eaters had more food to eat among the dead plant and animal matter than did the pure herbivores or carnivores. Birds, descended from a line of theropod dinosaurs, took over the niche left by the extinct Pterosaurs, and the age of the great mammals began. Deciduous plants thrived in the cooler climate; grasses created savannahs and prairies. The mammals expanded through adaptive radiation, some taking to the trees, like primates, lemurs, and the like, and some to the seas as cetaceans, such as whales.

The Natural History Museum of Los Angeles has an exhibit about the Age of Mammals. It headlines three forces in six words: *Continents Move, Climates Change, Mammals Evolve.*

Landmasses on earth are constantly moving. Mountains are uplifting, volcanoes are erupting, and fault lines are colliding or widening. The climate is constantly changing. Oceans warm and cool, glaciers melt and expand, the sun's heat increases and decreases. Most of these forces have nothing to do with us, are beyond our control, and have been happening for billions of years.

Today's climate, however, is warming so rapidly in comparison to the climate stories told by ancient ice cores and rocks that scientists say it is indisputable that human beings have accelerated the warming. How high can the carbon count's parts per million go before life for us is unsustainable?

The Age of Mammals is now the Age of Human Beings, or the "Anthropocene Age," as Paul Crutzen defines it today. Whether that age began with the first atom bomb detonations and the rise of plastics in the mid-twentieth century or the first use of fertilizers in the earlier part of the twentieth is negligible in the long scheme of things. We have made a difference. We have been making a difference since we discovered fire, made superior arrowheads, and walked out of Africa.

There is no mammal that can compare with human beings. We make things: symphonies and clarinets, wheat fields and bread, hospitals and MRIs, rockets and computers. And we make things to kill:

pesticides, herbicides, dynamite, guns, and bombs. We are the apex killer, the mammal that kills for revenge, for food, for land, for ideas, and for love. And for fun. No other creature is so versatile in killing.

We have conquered the earth. But the earth is resilient. She has been around for four billion years, changing with each new onslaught, adapting to the vicissitudes of space and time, surviving chaos. The earth will survive. It is we humans who will not. And we will take down many living creatures with us. But not all. Some will survive to evolve and begin again.

This is the most likely scenario as we continue down the path we are on. None of us today will be alive to see the end hundreds or thousands of years from now. Extinction takes time. It is a process of loss until the last of a species blinks out. No one really knows who shot the last Passenger Pigeon more than a hundred years ago; it took a while for everyone to conclude that they hadn't seen one anywhere for many years. No one has seen an Ivory-billed Woodpecker for decades, but there is still a glimmer of hope that it may roam the swamps of Arkansas or Florida. Until then it is not formally "extinct."

The most promising scenario is that we wake up and fix the amount of carbon we're emitting, which threatens to trap atmospheric heat, that we remove the particulates of plastic and chemicals that are killing plants and animals, ourselves included, and that we enter into a nurturing relationship with our planet, not an abusive one. It's like beginning a twelve-step program, changing ourselves for ourselves and others we depend on. Because this is always, and in the end, about people. About saving humans. Not animals or plants or water or land or air. The earth will take care of them if we take care of the earth.

When you fall hard for something, really fall in love, as I did with birds, the desire to protect it comes naturally. My involvement with the Wildlife Conservation Society began in the mid-1980s when Alan Rabinowitz introduced me to William Conway, its director, believing that my passion for animals translated well to conservation work. Dr. Conway agreed, appointing me readily to the Conservation Committee. My prior nonprofit experience had been with arts organizations and social-welfare groups, always lively meetings of people reflecting their constituency.

The Bronx Zoo was a special place for me. Whenever my family motored to New York from our suburban Boston home, my siblings and I were taken to the zoo. It was the biggest, the grandest, and the most beautiful of all. It was here that I first encountered a Galápagos Tortoise, an Orangutan from distant Borneo, and, best of all, the odd Duck-billed Platypus from Australia, an egg-laying mammal with webbed feet, which delighted my medical father no end. To be named to the Conservation Committee of the prestigious New York Zoological Society, as it was still called then, was humbling.

I nervously attended my first meeting in the stately administration building across from the Sea Lion pool. A dozen of us sat around a huge mahogany table, sunlight spilling through mullioned windows onto the polished surface. It was a quiet group. My fellow committee members were from New York's founding families; their grandfathers created the great cultural institutions of the city—the families Phipps, Haupt, Astor, Rockefeller, Pierrepont, and Frick.

George Schaller, on staff since 1959 for his seminal work on Mountain Gorillas in the Congo, gave a presentation. Dr. Schaller showed us slides of his current work with Pandas in China and of the glorious forests of bamboo where the Pandas lived. He spoke softly and told of the trials these creatures faced as their habitats were encroached upon. I was overcome by the story, by the stunning photos, and by the dedication of this extraordinary man. When he finished I burst into applause as I might for a brilliant Broadway actor. There was utter silence at the table, as if I had breached an unspoken rule of conduct. Dr. Schaller nodded slightly toward me and quietly left the room. The committee meeting continued with little discussion and then was adjourned, leaving me dazed and confused.

I felt like an outlier in the inner sanctum of New York. As the first actress on an NYZS committee, I was clearly of a different cut. It reminded me of the time in Ireland when a fellow actress and I tried to rent a car and were told in no uncertain terms, "We don't rent to theatricals."

Bill Conway put me at ease. He is the most genteel of all gentlemen, but I think he liked my rowdy enthusiasm and the ideas I subsequently brought to the table. Several years later I was appointed a trustee of the great institution.

The Wildlife Conservation Society was the spiritual home I had always dreamed of, giving larger meaning to my life. A thousand scientists in the field in more than fifty countries researched animals, finding ways to save them. I spent countless hours in WCS's zoos in Queens, Brooklyn, and Central Park, and in the Coney Island aquarium, becoming acquainted with the men and women who work behind the scenes as keepers and geneticists. I began to know the animals and was especially drawn to the new Bird House, which Conway helped create. I began to travel, first as a member of the Conservation Committee and later as a trustee, visiting biologists around the globe. I became friends with scientists whose stories surpassed the wildest scenarios of film and theater. My focus was shifting from the imaginary world to the wild and precious real one.

10

Peru

They seemed to appear out of the sky, bodies cresting the ridge in bursts of fuchsia, reds, and pinks, the children tumbling ahead down the tussock grass, barefoot and joyful in their race to greet us. The air was chilly at fourteen thousand feet in the Andes. My breath was caught; my heart was beating fast as my three colleagues and I labored up the mountain slope. The children jumped to a halt and waited for their elders to reach us, their dark eyes intent and giddy. The men arrived, shook our hands, and nodded, smiling.

Two of them in Western dress were the leaders of the community and of the project we came to realize. We were delivering five thousand polylepis trees to be planted by the community of Abra Malaga that day high in the Cordillera Vilcanota of southeastern Peru. This was a joint endeavor of the American Bird Conservancy, of which I was an early board member, and ECOAN, an ecological organization of the Andes. There were no trees to be seen at this altitude, only precipitous peaks in endless succession to the misty horizon and a small glacier nestled in a canyon high above. The village where these people lived was a sheer drop two thousand feet below, a cluster of tiny houses along a pale green river valley. It took them several hours to climb to where we were.

The existence of these indigenous Peruvians is precarious. Something drove them centuries ago into this inaccessible mountain terrain. They are descended from the original Incas who built Machu Picchu and other great citadels close by, and war must have been a constant for their ancestors. Today they barely thrive on the crops they cultivate: barley, oats, and tubers—mostly potatoes, of which there are more than a hundred varieties. But the winters are cold and they have denuded the hills of all burnable growth, especially the stunted, scraggly polylepis, the only tree that dares attempt colonization on this montane barren.

Although it is great to be credited with altruism for presenting this bountiful gift of baby trees for future firewood, we had an ulterior motive. The Royal Cinclodes, a rare bird of the Andes, faces an even more precarious existence than the people. There are only about 250 of them left in the world and they choose to live here and in another equally bleak region of Bolivia. They depend on the polylepis tree for shelter, for nesting, and even for the insects on the bark. With no trees they are doomed to extinction. We were riding in like the Lone Ranger to save the day.

A light rain began to fall, just a brief shower from a passing cloud. Bright blue plastic ponchos appeared and covered the people for the twenty minutes it took the wind to push the mist along and expose the sun again. Against the ocher of the grass the women and children looked like exotic tropical birds in their reds, pinks, and russets. These Peruvian women are master weavers, some of the greatest in the world, entwining tribal patterns in their knee-length skirts, shawls, and blankets, weaving symbols and codes into the fabric.

Babies bobbed safely on their mothers' backs, wrapped in intricately woven mantas. Every mature woman had a hat, a disc of cloth perched jauntily on her head, decorated with little red balls and buttons and safety pins of different sizes, and held tautly by a chin strap. It was cold, but some feet were bare or in flip-flops, since they preferred climbing this way. They were beautiful to look at, these women with their long black braids.

Two packhorses made the twenty-minute trip back and forth to the road all afternoon transporting the seedlings. The men scattered over the hillside digging holes while the women and older boys followed behind planting the tiny trees. I plunked myself down on a blanket next to one of the little girls given the chore of minding their younger siblings. We smiled at each other and giggled with the toddlers in universal play. I could not bring myself to lie down at the edge of the precipitous cliff, dangling my arms, as most of the girls were doing. One wrong move would mean sure death—a lack of confidence they clearly did not share.

Tino Aucca was heading the operation for ECOAN and Venezuelan Hugo Arnal for the American Bird Conservancy, as well as George Fenwick, ABC's president. It was Tino's idea to deliver the polylepis seedlings to these high mountains. He was born and raised in Cusco, the historic capital of Peru and the Incan Empire. With his strong aquiline nose and high forehead Tino could have stepped out of a gold medallion forged by his ancestors. As a boy he was drawn to animals and took the unusual step for his indigenous people of going to university, where he excelled in biology. At one point he thought his future would be more assured if he changed his name to Gonzalez, but his grandfather told him that their name, Aucca Chutas, was a very old Incan name meaning the "family of warriors" who protected the Inca, or king. Tino now keeps all things Indian dear to him.

He had taken us to Machu Picchu by train from Cusco, a breathtaking ride through steep mountainsides along a tumultuous river. I saw a male Torrent Duck in the rushing water, its striking black and white head and red bill bobbing through the foam. This is my favorite duck in the world; I am in awe of its skill, even courage, in navigating the weight and power of the torrent. Like a fish it streamlines through the water, stopping at eddies, where it feeds on mollusks, snails, and aquatic insects. Then it pops out onto a rock and preens and shakes before the next dive into the maelstrom. Competition for invertebrates from introduced trout and the damming of mountain rivers in the Andes is causing a steady slow decline of the Torrent Duck.

We die by bits. Things are taken away from us one by one until there is a failure to thrive. So it is with all species. Extinction is the natural order of things as new species evolve, and others like human beings dominate. But it is also the nature of human beings to arrest beauty, to hold on to it if only for a moment, a year, or a lifetime longer. We seek to preserve things of beauty if we can, for as long as we can, and, in the case of artists, for millennia.

How does one describe Machu Picchu? This isn't just a fortress, or a temple to the moon, the sun, or the soaring Andean Condor; this is a work of art on such an elaborate and massive scale that most other world architecture pales in comparison. One enters from the valley below or from above via the Incan trail at eight thousand feet; either way the reveal is astonishing. If you find yourself there on a cloudless day, as we did, the citadel is etched against the sky and held in the saddle between two sacred peaks that spear the blue. It is the stones that inspire us. As huge as those the Etruscans hauled into place, these Incan boulders by contrast have been molded with corners and curves into trapezoids and rectangles of different sizes, then hoisted snugly together with no mortar. The granite can look very hard and somehow soft as a pillow with gently beveled edges. Aqueducts proceed one on top of another in a zigzag of seventeen baths splishing down the mountain. There are stones that mark solstices and stones that act as a maquette for the sacred mountain in the distance. There is no part of Machu Picchu that has not been created with an artist's eye. To mark the day, a sleek Aplomado Falcon pierced the sky above us with a sharp wail, as his kind always has.

Tino also took us to the plains of Nazca, near the coast of southwestern Peru. In a Piper Cub hundreds of feet above the desert we saw vast numbers of lines etched in the pale soil and huge geometric shapes of animals, birds, and men dotting it for miles around. My favorite was a hummingbird, clearly defined by its long beak and wings. Walking the body of the bird on the ground later I found it impossible to define it from the trenches and dark rocks that made its outline. The shape was clear only from the air. Was this a shout-out to gods in celebration of the beauty of the bird? Or drawn for appeasement or a prayer for rain? There are lots of theories about the meaning of the Nazca lines, but no one really knows the truth, except that the artists lived around 700 AD and they clearly wanted to leave something of lasting value.

Tino wants to leave a legacy also, something lasting for his country: the protection of its indigenous people and its wildlife. We spent a week together along with a few others exploring sites that ABC and ECOAN seek to preserve for endangered birds. We saw the rare White-winged Guan in the dry forest of Limones, passing through dusty villages on the way, scattering goats and an ancient breed of hairless dogs in our wake. A pair of the birds perched on a cliff face hundreds of feet distant, but we could make out their slender turkey-like features with the white primary feathers as diagnostic. I have eaten guan in Belize, although not this critically endangered species, and knew that the tasty flesh sustained people throughout Central and South America. It was no wonder this White-winged Guan was down to about 350 individuals. Organizations like ECOAN are campaigning to save the birds, but it is a tough war against hunger.

In the Alta Mayo of the northeast our van negotiated treacherous mountain curves, braking often as we tumbled out to catch sight of a noisy flock of parrots or of raptors riding the thermals. Peru vies with Colombia and Ecuador for the most birds in the world, with close to 1,900 species. There are 135 different kinds of hummingbirds alone and in a little farming village that sits on a rise above quiet Lake Pomacochas at 7,500 feet, we saw one of the most spectacular.

This is the home of the Marvellous Spatuletail, an exotic if there ever was one. His territory, in the middle of agricultural fields of maize and wheat, is a brushy outpost of flowering plants and shrubs. Looking down into the tangle from a dozen feet above, we waited and waited for the tiny marvel to appear. Suddenly there he was, zipping by so fast we caught barely a glimpse. But the tail feathers were unmistakable—twelve-inch-long flexible wires of white with black dots at the tip, undulating like seaweed fronds as he probed each bright flower for nectar. Yes, indeed—"marvellous"!

Birding is a peculiar but deeply satisfying occupation—perhaps not an occupation at all but an obsession. We rarely get the perfect look at a bird that field guides or apps give us, but just being in the zone, the quiet and totally focused state of watchfulness that is required, is reward in and of itself. Our days in Peru were packed with birding in remote areas, checking off hundreds of species on our lists, prioritizing what to protect, and searching in vain for the most reclusive, like the tiny Long-whiskered Owlet, recently filmed for the first time in the cloud forest of Abra Patricia.

We were lucky that afternoon high up in the mountains of Peru, where scant representatives of remaining polylepis bent low to the wind in exhaustion. Suddenly someone called out, "The bird!," and there he was, several hundred feet above us. In the circle of my binoculars I saw a thrushlike chocolate-brown back, a fleeting white cheek with a dark eyeline, a fluttering among the small trees before he dashed over the horizon and was gone: the Royal Cinclodes. You have to admire a bird that can live at fourteen thousand feet. Some of the Corvidae family, crows and ravens, hang out there, and some raptors such as Lammergeiers in the Old World, but few passerines like the bird we just saw. It was enough to bear witness to its presence. And the feeling was rapturous. Spotting a bird you've been after for hours, or days, and finding it in a shaft of sunlight or hearing it sing in the forest is like peering into a bit of heaven. The image is indelible and birders have catalogs of mind photos they carry forever marking these sightings.

If not for birders, the Royal Cinclodes would have been relegated to the annals of "once was." Most people wouldn't care; it's not a bird they would ever see. What good is it in the world anyway? It does not feed anyone but a passing raptor, perhaps, and if it serves any purpose, it is difficult to know what. Its purpose is not ours to know. It exists, and it is an extremely rare species that any knowing birder wants on her list. So by serving the birders' specialized passion, the people of Abra Malaga benefit, and the Royal Cinclodes is spared an early demise.

The afternoon cooled as the sun lowered in the sky on the glacial ridge. The planting was done and the people gathered together on a small plateau and asked us to join them. The men formed a

*Looking goofy but touched by the gift of little hats
in the Peruvian Andes, 2004*

circle while the women and children kept their distance. But not us; Carol, a photographer, and I were asked to stay. The village leader began to speak in Quechua while Tino and Hugo stood by, listening. Tino smiled, nodded, and shook hands with the leader, proud of the project he had conceived for his fellow native people. Then two knitted caps with earflaps and long tassels appeared, the kind worn by the children. Tino beckoned us forward, and Carol and I shyly moved into the circle. I was embarrassed by our raggle-taggle hiking clothes, so drab in the midst of their kaleidoscope of color. They placed the caps on our heads and we were clearly thanked for our contribution to the day. This was unexpected, and tears flowed down my cheeks as I bowed my head trying to hide them. I fumbled with the brightly patterned cap and tied the woven straps under my chin as best I could. It was so small it perched like a top feather on my head. I laughed and then everyone was laughing as someone snapped a picture so the moment was caught forever.

It was not lost on me that the way of life of these people is on the verge of extinction. As with the Royal Cinclodes, the end is near, but for the people it will mean only a change of clothes, and perhaps language, education, better health, and shelter. The trees are planted.

Now, years later as I write this, a half million polylepis seedlings have been distributed and are recolonizing parts of the high Andes. Dozens of communities have received thousands of fuel-efficient stoves and badly needed health care as well. One of the rarest birds in the world has been given a reprieve. Human beings intervened and gave the bird back the habitat it needed to survive, for the time being. Eventually the warming climate will push the Royal Cinclodes farther and farther up the mountain until there is nowhere else to go, and then one day the last of its kind will rise into the sky forever.

11

Papua New Guinea

In Shakespeare's dark Scottish play, the tragic hero Macbeth has been given the prophecy that nothing can harm him "till Birnham forest come to Dunsinane." Knowing that forests do not move, Macbeth is lulled into believing his life is safe. But his enemy Malcolm has his soldiers camouflage themselves with tree branches as they storm the castle hill. So Birnham forest comes to Dunsinane after all, and Macbeth is vanquished.

I did not believe we were going to be vanquished by the phalanx of grass-covered New Guineans advancing up the landing strip in the Huon Peninsula, but the unexpected nature of the vision before me brought the quote to mind. Only in the theater had I encountered fantastic happenings like this. But this was real, in secluded mountains on the other half of the world. Everything about New Guinea was surreal—from the people and their elaborate decorations, to the Birds-of-Paradise, to the teddy-bear kangaroos that lived in trees.

The hidden men shook their grassy skirts, made a guttural chorus of sounds and high-pitched squeals, beat long bongo-like drums, looking first one way and then the other as they came closer and closer, then stopped abruptly. Their faces were obscured by long fronds on their rattan headdresses, adding mystery to the moment. This was our formal welcome to the village.

The biologists Bruce Beehler and Lisa Dabek invited three of us to join them in Papua New Guinea at their field sites and to witness conservation efforts. This was the inaugural visit of Conservation International's Sojourns program, where Bruce worked. Bruce is a self-described "bird man," a leading ornithologist in the study of Birds-of-Paradise and Bowerbirds and the coauthor of *The Birds of New Guinea*. In 2005 he and a team of eleven scientists did a biodiversity blitz in the secluded Foja Mountains of western New Guinea, photographing and describing many birds, frogs, insects, and mammals for the first time. He discovered a bird new to science, the Wattled Smoky Honeyeater. This and the elaborate courting behavior of the Bowerbirds and the Birds-of-Paradise led to *60 Minutes* filming Bruce there in 2007, which made him a celebrity. I first met this delightful ornithologist at a benefit for a conservation organization. Bruce has a modest nature when it comes to his own remarkable accomplishments and an easy sense of humor. All it took was for me to say I would love to see Birds-of-Paradise and two years later we were on our way.

Lisa Dabek grew up an asthmatic child in New York City who fell in love with animals even though she was allergic to their fur. She became senior conservation scientist at the Woodland Park Zoo in Seattle, where she first encountered the rare Tree Kangaroo that became her life's work. Lisa's study site on the Huon Peninsula overlapped a transect where Bruce and other scientists were conducting long-term biodiversity studies on climate change.

We flew in a single-engine plane to the village of Yawan, four thousand feet below the site, while we waited for a helicopter to take us all the way up the mountain. The Huon Peninsula is in the eastern half of New Guinea, part of the independent nation of Papua New Guinea, while the western half is part of Indonesia. New Guinea is the largest tropical island in the world, with the highest mountains of any island, some reaching fifteen thousand feet. It is close to the equator and carpeted in green. Everything about it is big, bizarre, and beautiful.

The welcoming ceremony, called a sing-sing, moved down the grassy hill, past huts of bamboo thatch with borders of flowering shrubs as neat as in any U.S. suburb. The village was ringed by steep green mountains, a thin waterfall splashing its way down. People tossed flower petals on us as we entered the village. Leis of pungent marigolds were placed around our necks, and my companions and I were welcomed in Tok Pisin, or Pidgin, the lingua franca of New Guinea, where more than eight hundred languages are spoken. Lisa hugged the villagers and asked after them with a warm smile.

We sat on benches in the bright sun and watched a performance. Teenage boys, their chests, legs, and faces painted with white and gray mud, wearing hula-type grass skirts, huddled in a teepee of banana leaves and flowers, trembling. Suddenly they burst from it like a rainstorm, whirling around in different directions, until they tore the teepee down in a dramatic finale. The storm was over. There were brief speeches of welcome and we reciprocated with thanks for the ceremony. Then we were draped with tapas cloth and given a colorful bag, the native "bilum" worn across the chest by all New Guineans, young and old. We had arrived.

But the helicopter didn't. It couldn't get through the cloud cover, a common occurrence in PNG, where it rains a lot. We spent the afternoon visiting the tidy school, where boys and girls learned about animals and ventured a timid "My name is Winston . . ." We walked a narrow path to a neighboring village, logging our first birds and butterflies, and were surrounded by curious children in worn clothes from a "bale," the donations of Westerners like us. Back in Yawan, Annie, a village leader, provided us with a meal of rice, greens, taro, and sweet potato, and we bedded down in one of her little thatched huts on stilts. New Guineans are small people, and we five Americans towered over them by a good foot. In addition to Bruce, Lisa, and me, our group included Caroline Gabel, whose Shared Earth Foundation conserves wildlife, and George Meyer, a creator of the TV series *The Simpsons*. It took some doing to squeeze our sleeping bags and gear into the space, but I slept okay next to George, who at six feet six inches had to bend his knees to fit.

Around 3 a.m. I rose to go to the outhouse, which required stealthily getting my boots on, climbing down the ladder, and walking with my penlight on the slippery muddy path three hundred yards down to the two-holer. Insects and frogs filled the night air before I heard a deep escalating chant—a kind of rocking bass chorus that reminded me of Sioux Indian chants in North Dakota. Yawan is a Seventh-day Adventist village, but these were clearly not hymns, not at three in the morning.

PNG is nominally a Christian country, having been converted by waves of missionaries since the nineteenth century. But animism and a belief in spirits still thrive in pockets alongside Christianity. The Adventists brought peace to Yawan, as well as sobriety, health care, and education, but much of New Guinea has a deeply ingrained culture of violence. There are hundreds of tribes or clans spread over the mountains, the rivers, and the coasts, in isolated areas where footpaths and waterways are the only means of travel. Different cultures emerged, each with different languages, dress, and art. Marriage is not allowed within the clan, so wives come from elsewhere. Constant warfare over boundaries, wives, and pigs was waged for centuries. So were cannibalism and headhunting until the 1950s, and in some areas into the 1970s, when government programs and Christian practices put an end to it.

Yawan is a peaceful village, as are the other forty Lisa works with in this northwest sector of the Huon Peninsula. Twenty years of building trust with the people through a plan for economic development as well as environmental sustainability resulted in the Tree Kangaroo Conservation Program. It directly helped twenty-six teachers graduate and return to their villages to teach; it donated solar-powered lanterns to aid health care workers; and it initiated the production of tree-shaded coffee for the Seattle market. In return the villages have developed multiple land use plans to mitigate what is called the "careless use of resources," one of which is the endangered Matschie's Tree Kangaroo, which New Guineans love to eat but no longer hunt in its protected areas. Conservation is not possible without buy-in from the local people.

The helicopter arrived bright the next morning and we spiraled up the mountain to ten thousand feet and a glorious cloud forest.

The helipad was a cleared spot overlooking ravines to the east and west. Paw prints of wallabies dotted the earth, and birdsong filled the air.

In a cloud forest everything is wet. The footing was slippery, and mud smothered my hiking boots, but the riot of ferns and mosses, of orchids and bromeliads lacing the tree branches, was sumptuous. The Wasaunon field camp was above a little stream; a tarp-covered lean-to was our meal shack where we dined on crackers, soup, rice, and greens on place mats of palm leaves. The ten PNG fellows did everything with artistic flair. They put up individual tents for us and masked our latrine with a bower of ferns and flowers, as if for a Disney princess. I was glad it was so attractive on the outside, because inside the trench was so slippery it took infinite skill to squat.

At ten thousand feet and forty degrees you don't get the variety of night sounds you do at tropical elevations. But at dawn the songs of Spangled and Red-collared Honeyeaters rang out along with the din of cicadas. After a cup of rich New Guinea coffee and boxed cereal we all set out to find the Tree Kangaroos. Gabriel Porolak, Lisa's PNG assistant, led the way. Gabe had tracked "Trish" and her joey the day before by her radio collar, so he knew where she was. Still, the thickness of the vegetation and the height at which Tree Kangaroos live made the exact location difficult to pinpoint. The ground was spongy with mosses and saturated earth, while tree roots crisscrossed the forest floor. I had to take care not to trip as we made our way through the dense foliage of tree trunks, saplings, and tall ferns.

How one of the guys spotted Trish was beyond me. It took several minutes to find her with my binoculars, about forty feet up on a large branch of a tree. Her golden-brown tail hung down like a thick vine while her body was perfectly camouflaged in sunlit leaves where she was munching on a spray of orchids. Not far away was her joey, half her size, never before seen by Lisa. He was out of the pouch by ten months old, so she guessed he was about a year and still going to his mom to nurse. They become independent at eighteen months. There are fifteen subspecies of Tree Kangaroos, all evolved, along with terrestrial kangaroos and wallabies, from an ancient arboreal opossum. In drier Australia they became bigger and stayed on the ground, while in New Guinea they evolved into distinct subspecies in remote niches.

Every year the batteries need to be changed in the radio collar, and Trish was due. Her radio collar these past four years had taught Lisa and her staff a great deal about Tree Kangaroos—their range, courtship, what they eat, and the threats to their survival.

The area underneath the branch was cleared of anything that might hurt the animals on impact. A young man of fifteen elected to climb the tree to induce her to jump. He started up the trunk barefoot like a Hawaiian native going for coconuts. But this tree was so big he could not get his arms around it, and his prize was forty feet up. Trish climbed higher, her little joey following as the boy came closer. Every time she moved up, so did he. We watched in admiration as the boy put hand over hand, foot over foot, as assuredly as Spider-Man, and reached the branch below where she finally stopped sixty feet above us. He noisily beat the trunk and Trish moved to the end, arching the tip with her weight, while her joey sought to join her. He must have thought better of it because he suddenly jumped, flying through the air for the first time with true grace for a round ball of fur. He landed on the soft forest floor unharmed, and the men ran to grab him. Trish made the flight herself within minutes, twenty pounds of her landing without a scratch. One man grabbed her by the tail and others held her by her neck and legs. Gabe removed her old collar and replaced it with a new one. She was relatively calm, having been through this three times before; and she was perfectly beautiful. Her plush tawny-orange fur circled a white belly, and her big eyes gazed at us as she suffered the indignity of being held upright in a splayed position. Her long black nails curved at the end with sharp points, good grips for tree climbing. Her joey was immediately christened "George" by the New Guineans in honor of our colleague George Meyer, who was deeply touched. I stroked baby George's soft coat, touching his paws and his sweet gentle face with its pink nose. He was about the size of Paddington Bear— a huggable teddy.

The Huon Peninsula is the only area of the world where the Matschie's Tree Kangaroo lives, making it one of the more endangered mammals on earth. They are secretive and shy, living high in the canopy where mating is a delicate endeavor. A large arboreal python has a fondness for Tree Kangaroo, and they are prized meat for the native people as well. Hunting and loss of habitat are the main threats to the species.

Lisa has been educating the villagers since 1996 about this rare creature in their midst. What began as a research project evolved into an alliance of local people, scientists, and research institutions. There are forty-five villages and more than ten thousand people in the YUS region of the Huon Peninsula, named for the area's three main watersheds, Yopno, Uruwa, and Som. The clans own all the land, which is true of most of Papua New Guinea. They took charge of their own future by creating PNG's first conservation area, formally recognized by the government in 2009, giving the area the highest protection under the law. It means there will be no extraction of resources, including mining and logging. The villagers are still able to hunt the kangaroos sustainably in certain places but they are off-limits in 180,000 acres of protected areas. So far the villagers are honoring the rules. The Matschie's Tree Kangaroo is a source of pride to them and to their children, who learn about the animal in school. The staff of the Tree Kangaroo Conservation Program in the PNG office are all native men and women, which TKCP aids in further education. Gabriel Porolok completed his PhD at Cooke University in Australia and is now a teacher of biology in his homeland of Papua New Guinea, specializing in native flora and fauna. The Matschie's Tree Kangaroo is his signature species.

We spent the next day in our mountain aerie walking the climate change transect, which Bruce Beehler and fellow scientists are documenting over the course of many years. What will a warming climate mean for the animals and plants there? The path in this stunning primeval forest led us through ferns and mosses and abundant orchids, and past many species of trees, some 120 feet high, all of them new to me. Birds were the only visible life but hard to see against the green canopy. We glimpsed the Crested Berrypicker, different kinds of Honeyeaters, and lively Fantails, which flitted about with a flashy tail display.

There are no ferocious mammals in the Melanesian forests—a few boxing wallabies perhaps, but no felid or canid species, and no large bovids, ungulates, monkeys, or squirrels. Poisonous snakes, spiders, and insects are there, and even a poisonous bird. Some Pitohui bird species contain neurotoxic alkaloids similar to the Poison Dart Frog of the Amazon. The New Guineans have long known they are lousy to eat and have steered clear of them, but it took a Western scientist ingesting some of the poison on his hand to "discover" the unique bird.

The next day it poured, and the mild temperature plummeted. We read in our tents and played Scrabble, which George, with his quick wit, won. We had a discussion about conservation ethics. Some of us were concerned about the ongoing collection of species for science at a time when so many are threatened. Bruce defended

Catching up on birds in the bush plane, on our flight to Tari Gap, PNG, 2010

the practice of collecting a minimum of five individuals of a species; he felt science needed to know as much as possible in order to save them. I said that collecting needed to be better regulated, as the right hand often did not know what the left was doing and there were too many people from too many places taking too many individuals. Lisa did not think killing an animal for study was necessary anymore, that technology had changed the way we learn.

The flight to the Highlands took us right to Ambua Lodge, decidedly upscale after tenting in Huon. The buffet lunch was a feast of fresh fruits and salads. Our spacious huts overlooked an expansive valley below, where bush planes landed several times a day. We shared the lodge with field managers of the huge liquid nitrogen gas company that was building new roads for drilling, promising jobs and wealth for the country. How long, I wondered, before they penetrated the most impenetrable areas of New Guinea with their technology?

After lunch Bruce found a flowering tree attracting the Superb Bird-of-Paradise and the Brown Sicklebill. This is what we had come for: the Birds-of-Paradise, what Alfred Russel Wallace in 1869 called "one of the most beautiful and wonderful of all living things."

Wallace, the eccentric and brilliant biologist who published his theory of evolution at the same time as Charles Darwin but received virtually no attention for it, spent many months in the mosquito-infested Aru Islands southwest of New Guinea. He was the first Western scientist to see the Greater Birds-of-Paradise displaying in their "lek," their frothy orange tail feathers rising and falling as they twisted and turned in communal dance. He did not see the extensive variation in all thirty-nine species because travel did not allow him the privilege back then. But he saw enough to ponder the question: What was the origin of new species and of variation within species?

He knew that species went extinct. Fossils told him that after millions of years the end came for most species. How, then, did new species arise? He didn't buy the "special creation of God" dogma; he believed that new species were built on those that had come before. It was Darwin who used the phrase "natural selection," while Wallace's "gradual introduction" had yet to become "evolution" for both. A hundred years later the ornithologist Ernst Mayr was still deliberating: "Birds of paradise raise difficult questions, questions that penetrate to the very foundation of our biological theories. How can natural selection favor . . . the evolution of such conspicuously bizarre plumes and displays? How can it permit such 'absurd exaggerations' as one is almost tempted to call them? How can it happen that apparently closely related species and genera differ so drastically in their habits and colorations?"

Scientists today know that the "absurd exaggerations" of Birds-of-Paradise are due to sexual selection. "BOPs," as they are nicknamed, and Bowerbirds live in a monkey-free world where fruit is abundant. They have no competition for resources, so they don't spend an inordinate amount of time looking for food; this frees them up for courtship. The Bowerbirds create "courts"—mini-theaters, or bowers, where they build sets, erecting towers of sticks, grasses, or vines that they decorate with flowers, stones, plastic, berries, in fact anything colorful that they find, even Coke cans. Then they engage in some performance art on the stage for the females. They dance, they present gifts such as a bright red berry, or even a ring they've found, and they vocalize, mimicking animals, trucks, and human beings or running water. They are great at it. Some of the males even arrange stones in front of their bower to create perspective, the largest in front to the smallest in back, rivaling the work of some of our best Broadway set designers. This optical illusion seems to win the hearts of many females, and it's not surprising, is it? There is a big "wow" factor there. Great artists, like great athletes, attract ladies the world over.

The Birds-of-Paradise work with their own bodies to entice females. Some of the species have evolved remarkable coloration, which they shake, run up and down poles, or hang upside down to show to best advantage. We saw the Raggiana Bird-of-Paradise the first day in Varirata National Park not far from PNG's capital, Port Moresby. The lek, or communal courtship area, was about thirty feet high in dense trees. About twenty males were bouncing around dipping their yellow heads to better show their copious apricot-colored plumes to the females gazing from branches above like so many matinee ladies in the mezzanine. The males squawked and jockeyed for position while the females flew in and out.

The King Bird-of-Paradise has purple legs, silky red feathers on his head and neck and lower body, a ruff of beige and teal-blue feathers below his neck, a white chest, and two green fiddlehead discs at the end of long tail wires, which he can raise or lower at will. You get the idea. These colorations are outlandish, and like nothing seen anywhere else. Perhaps half of the thirty-nine species have these remarkable costumes, which they strut to entice. The other half are what Tim Laman and Edwin Scholes call "shape-shifters."

Tim and Ed spent eight years in New Guinea photographing and describing all thirty-nine species. Their book, *Birds of Paradise: Revealing the World's Most Extraordinary Birds,* is a remarkable achievement, highlighted by luscious photography and compelling insights, one of which is the descriptive term "shape-shifters."

Bruce pointed out the Superb Bird-of-Paradise to us as it was bobbing about in panicles of white flowers. It was a female, not the male, who does the shape-shifting, but wonderful to see anyway and to know we were in their neighborhood. The male is all black except for a turquoise breast sheaf of feathers, which in display jut out from either side like the wings of a blue angel. When he opens his mouth to sing, the inner lining is lemon yellow. The shape-shift comes when the female is around. He lifts his entire cape of back and shoulder feathers and flips them over his head, creating a perfect half circle. With his head down, showing two tiny blue feathers like eyes above his ordinary black ones, his turquoise feathers spread like wings symmetrically to either side in front of the black cape; he looks like a huge smiley face sitting on a log. Tim and Ed made it possible to see this on film because it is rarely seen in the wild, and only patient biologists and their mentors Bruce Beehler and Russell Mittermeir have ever had the chance.

We saw a few more female BOPs that afternoon before the rain came down in torrents. I went to bed early, a cookie and a roll folded in a cloth napkin on my bedside table for a snack, only to find every crumb gone at 5 a.m. Little rat fingers had neatly opened the napkin and made off with the goods.

The Tari Gap, at seven thousand feet, is a great place to see the BOPs that live at this altitude and higher. We stood on the new road built by the liquid nitrogen gas folks and listened to all the morning birds tuning up. A male King of Saxony sat atop a tall tree stump rattling away. A Robin-sized bird of yellow, black, and white, he has two plumes twice the length of his body protruding from either side of his head. These aren't ordinary feathers; they are more like wires with hard fish scales attached like tabs all along them. He can move these feathers any which way, much as Uncle Charlie can wiggle his ears. These antennae are highly prized by the New Guineans.

We saw more Brown Sicklebills, another shape-shifter, with their long curved beaks and their loud rat-a-tat-tat like an AK-47. But the most remarkable sight for me that day was the Ribbon-tailed Astrapia, which lives only in this one area of the world and has evolved as it has only because it has few predators, besides man. From the ridge I looked out over the endless green forest to the spine of the next mountain far in the distance. I thought someone had released a party balloon, because over the ravine two long ribbons floated rhythmically by in the sky. At first I didn't see the dark body of the bird against the green, but when I did, the sight of it flying with yard-long tail feathers, three times the size of the bird and the longest of any bird in the world, was simply astonishing.

While the birds of New Guinea are unrivaled in the world for decoration and transformation, the same can be said of the people of New Guinea. Surely the birds influenced the costumes, dances, and art of the clans. It probably began in the Stone Age. Some fellow picked up a feather or two and put it in his hair, maybe with a few ferns and a peekaboo skirt of grass. It made a young lady smile. Add a few twists and turns, a couple of dips, and theater was born. The Birds-of-Paradise must have given the New Guineans lots of ideas, because there is nothing in nature as imaginative as their colorful displays and their shape-shifting dances. Every clan has their own unique way of decorating themselves, moving and vocalizing. This came about because of their isolation from each other. Today as many as seventy-five clans come together for an annual sing-sing festival at Mount Hagan in the Highlands, an event where tourists are welcome.

Costumes and makeup are a big part of my life. I have spent countless hours in my dressing room backstage applying line and color, seeking transformation. When the arch of my eyebrows and the rouge on my cheeks is just so, when the cut of my hair and the height of my heel is just right, when I don my petticoats, my cape, and my hat, I am finally in character and I am ready to make my entrance. The external change in me influences the internal. It is always transformation I am seeking—revealing parts of myself I didn't know existed, becoming someone else. It is liberating to experience oneself as another being, and it is also revelatory. Becoming someone else is becoming everyman and everywoman. The universality of being human is found within each of us when we engage in shape-shifting . . . and perhaps of being animal too, a universality of all living things.

We went to meet the Huli Wigmen, the famous Highlanders of the Huli clan who use their own hair to create fabulous wigs, which reminded me of the Afros my black friends sported in the late 1960s. Those that choose to go to wig school grow their hair for eighteen months, spending nights sleeping with their neck on a block so it doesn't get mashed. Then they cut it and fashion wigs for day use and for ceremonies. These headdresses last a long time and can be bought by others in the clan for about $600.

The wigmen performed a sing-sing for us. The best part was seeing them paint their faces before the event. It was a sneak peek backstage in the sunlight, as there was no dressing room per se. They had little bits of broken mirror and pots of bright red and yellow paint and small brushes. The process was the same as the half hour before any New York curtain. The chief's concentration was fixed as he glimpsed his face in the mirror; he applied his makeup masterfully, yellow covering his entire face and then a red line down the middle

*A New Guinea clansman making up
"backstage" before the dance*

of his nose, and two across his cheeks. At the end he pushed a long thin twig through a hole in his septum and the transformation was complete.

Brief squares of cloth covered their fronts and backs, and like Scotsmen over their kilts, they wore a kind of sporran with bits of animal fur hanging from their waists, which bobbed up and down as they hopped in unison. Their necks were ringed with shells and beads and their noses and ears were pierced with bones, feathers, or stems. They dipped and swayed like the birds, and in homage to them, their headdresses of hair were stitched with perfectly placed feathers of Raggiana, King of Saxony, and my beloved Ribbon-tailed Astrapia. They killed the birds for their feathers, but the care they gave the feathers made them last for years and has kept the birds from being decimated. There are species of BOPs and Bowerbirds that are on the threatened list today but due more to the new gas highway, habitat loss because of development, and the disintegration of traditional culture.

When we were in PNG an Irish company had recently erected towers on the high peaks and given out cell phones for four-month trials. The clans traded in the currency of pigs and had no written language, but several of the men mastered the buttons in a matter of hours and were chatting to each other across the village with dexterity and frequency. Perhaps the human brain is hardwired for this kind of thing. That little wonder in the palm of their hand suddenly ended a way of life that had prevailed for thousands of years. Just one cell phone in each village, conveying information, eliminated the need to walk two miles to the next village, something that had been a daily ritual. I wondered how their social life would change. I wondered how they were going to pay for the cell phones when the bill came due—in pigs? And what was so important that they were talking on them so incessantly? But then I wonder the same thing of my texting granddaughters, who are wearing out their thumbs with the need to communicate. Human beings are human beings the world over, and a new device has the same power to enthrall everywhere. The New Guineans will never be the same, and neither will we, with each new technological advance that enters our lives.

It is by increments that cultural traditions dissolve. One day someone dies and a recipe is gone forever, or a language, or a dance, or the technique for weaving a Huli wig. The new generation is not as interested anymore in learning from their elders. They flee to Port Moresby and listen to Western music and wear jeans and get into trouble, like kids everywhere. But the territorial culture of violence follows them. In the Highlands the violence never went away; "an eye for an eye" still rules. Debts are paid in pigs, wives, and the local currency, kina, in that order. And when trade doesn't work out, things get bloody. We met a young British woman working for Doctors Without Borders who said she operated daily on amputed fingers, hands, and even arms and legs as a result of clan warfare in the Highlands.

Still, in the clan there is a structure that is adhered to. If the Birds-of-Paradise need to be protected for their feathers it is likely they will be by a clan that preserves its traditional sing-sings. The young men in Port Moresby don't care anymore, and things fall apart. It is a violent city with a high rate of crime and homicide. There are many cities in the world that tell the same story, cities in Central America, Indonesia, the Mideast, eastern Europe, and the United

States, where the innocent have been killed by disenfranchised mass murderers. Violence is transformative—certainly for the victims, but also for the perpetrator. With a deadly weapon he becomes powerful, his persona changes, and he becomes an actor in his own play. Urban gangs or rural terrorists become companies of players writing their own destiny. They are not connected to the larger world because they have created their own. They are shape-shifters, dangerous ones.

My trip to Papua New Guinea began with shape-shifting birds and ended with shape-shifting men. There is a correlation and a difference. The birds transform themselves to be the apex male, but it is the female bird that decides which is the worthiest. Men transform themselves for the same reason, but also to be the apex predator in society, a rite that becomes violent.

When we lose societal traditions of transformation, communal rituals, and performance, we lose the extension of human experience. We turn inward, unable to express all the identities within us. We live vicariously through the stories of others on TV or in movies or books. We thirst for stories and tune in to reality shows, or gossip, or Google, or Twitter, or constant communication with friends to get the latest event. Ours is a virtual world because the real one has become stunted.

A friend took her two daughters to see an exhibition at the American Museum of Natural History in New York. The exhibit included two rare Galápagos Tortoises, some of the oldest creatures in the world. The girls walked right by them to see the video playing on a nearby wall. Their mother was shocked and explained how extraordinary the tortoises were. Her daughters said that it was unnecessary to bring the tortoises all this way when a video told the story better. Perhaps they were right regarding the tortoises' welfare, but the virtual world diminished the girls' experience.

Bruce and Lisa stress that there is no conservation without people. But without people who care about wildlife, there will be no conserving it. New Guinea is one of the most remarkable islands on earth, with unique plants, animals, and human beings. There are hundreds of spoken languages but none written, and most are spoken by fewer than a thousand people each. The six thousand languages in the world are declining at an alarming rate, and New Guinea is the country with the most languages per capita. New Guineans are walking libraries, lives to be celebrated for their infinite variety of artistic expression and for their profound knowledge of the land and the creatures that live there. Cultural diversity and biodiversity go hand in hand; they thrive together and die together. It is fortunate that the clans own 98 percent of Papua New Guinea, because it is local people who will be invested in protecting what they own. It is the young men who need to be wooed back. Perhaps the marriage of technology and wilderness can work in the long run. Maybe the Irish had it right after all when they introduced cell phones.

12

Hawaii

The east coast of Haleakala is not a well-trod route, but I was on a quest. I had read that the White-tailed Tropicbird lays a single egg on a ledge of a volcano, and I thought my best shot to see one would be to drive to Hana for the night, then continue the next day past Seven Pools and onto the dirt road that traversed the east side of Maui. This was the kind of thing I would do on my days off when shooting movies.

Kris Kristofferson, Madeleine Stowe, José Ferrer, and I were in a two-part miniseries called *Blood & Orchids* in 1985. It was a compelling crime drama based on a true story of racism in Honolulu in the 1930s. I played the bad gal while Kris played the detective who ultimately takes me down. We shot in glorious locations around the island of Oahu, and when the weekends came, I flew to one of the other islands to see the birds there.

My life as a birder actually began when I started traveling for work in movies or plays. I was looking for a hobby that didn't require carting around a lot of gear, and birding seemed perfect. Wherever I went there would be new species, and all I needed was a field guide and a pair of binoculars. From the very beginning back in the early 1970s, I was hooked. At first I would contact birders in the area through Audubon or friends and ask if they would take me around to good local spots, but often an erratic shooting schedule would cause me to cancel the appointment, and so as time went by, I began to rely more and more on myself. This was a slow way of learning but more intensive. No one was telling me which bird I was seeing, I had to learn it myself with the help of the field guide; my observations became more acute.

Every new location became an adventure, and I began to know the parks of Chicago, Toronto, Boston, and Los Angeles as well as I knew Central Park or my own backyard north of the city. And I explored parts of Spain, Germany, England, Australia, and Hungary as most tourists never did. Everywhere I went, people were helpful and sometimes amused at this lone woman walking in out-of-the-way places.

I never had a problem, except once. Across the freeway from my twenty-first-floor hotel room in Fort Lauderdale, Florida, I spied a waterway beneath an overpass and couldn't resist checking it out. Criss-crossing the four-lane highway, I slid down a twenty-foot embankment, scattered some Mallards paddling about, and then caught sight of a real prize: a Limpkin. This large water bird was poking its impressive beak into a snail shell down by the culvert. I squatted down, binoculars to my eyes, fixated at the sight of this uncommon bird, when a creepy feeling made me turn around. There just thirty feet behind me was a scruffy man, sitting on the muddy bank and staring intently. Then I saw the knife in his hand, and wasted no time scrambling up the bank and dodging cars as I raced toward my hotel. I looked back, horrified to see him still in pursuit through the traffic. Intent only on getting to the safety of my room I tore through the huge lobby, never thinking to alert the staff to the creep. I made it to my room, hoping he had not seen the twenty-first-floor elevator button I pushed, and I stayed there the rest of the afternoon.

Incidents like this, and worse, do happen to birders, but usually in remote areas of the world, not in a bustling city in Florida. Phoebe Snetsinger, a hero of mine and one of the most famous birders of all time, was raped in Papua New Guinea when she was on a mission to see more birds than anyone else in the world. Phoebe was indomitable; she took the incident in stride and became the all-time champion lister before her death in 1999, having seen 8,400 species, or 85 percent of all the named birds on earth.

My mission on the back side of Haleakala in 1985 paled in comparison to the great Phoebe, but I had a deep desire to walk the ocean cliffs of Maui in hopes of seeing the White-tailed Tropicbird. After an exhilarating drive in rain the night before on the treacherous winding road to Hana, I thought the dirt lane past Oheo Gulch in broad daylight would be a cinch. I had a refreshing dip in one of the seven sacred pools, dunking my head under a chilly waterfall. Then I picked a bagful of soft ripe guavas before putting my car in low gear and driving on. The bumps were significant, jouncing me up and down like croupier's dice, making me rethink my contract with Avis. But I made it to Kipahulu Point Park, just as a cool mist blew onto the slopes. Charles Lindbergh lies peacefully in a grave nearby, overlooking the unbroken ocean he must have flown over dozens of times. It is a lonely place but not desolate, befitting a man whose life had many sorrows.

After walking the trail for a while I sat on some rocks, ate some guavas, and looked out past the waves crashing far below me. A few Frigatebirds, looking like small Pterodactyls, plied the air in easy circles. My bird would be more direct, headed toward something if she wasn't plunge-diving from fifty feet up into the Pacific for fish. Somewhere high above I thought there must be a perfect rock crevice for her nest.

I do not pray hard for things, but I do envision them, and I had a vision in my mind of the graceful white bird winging her way in through the ocean clouds and then up over my head toward the crater above. The mist was enveloping me now, the air cooling, and I thought the drive back might take double the time. I was called for shooting in Oahu the next day and couldn't miss my flight.

One last swing of the sky through my binoculars, clouds crowding the lens, and then a tiny speck, not dark like the Frigate, but blending into the white of the fog, was riding the air currents far above the waves. This bird was moving with purpose; she closed the distance in my lens until bingo, there she was, fully revealed, my White-tailed Tropicbird. What a buoyant bird, graceful and sleek as any aloft, a line of angelic white against the fog as she headed toward the volcano, her long tail streaming behind her like a beacon of light! Did I call her in with my vision? It has happened to me before. I have sat where a bird is likely to be at a given time and waited, envisioning its appearance. Usually nothing happens, but when it does, there is a deep stirring of ancestral experience, that I am part of something far greater than myself in the juxtaposition of bird and me in time.

The Hawaiian Islands are some of the most remote on earth, a distance of 2,400 miles from the West Coast of the United States, and from Asia. They are young and volcanic, and some of the volcanoes, like Kilauea on the Big Island, are still active. Like all remote islands, the Hawaiian Islands were colonized in the very beginning by the three Ws—wind, waves, and wings—which carried seed and spore across the vast ocean. Birds probably brought most of the seed that took hold in the fine Hawaiian lava soil; it came in their gut, and on their wings or feet. Some birds are remarkable long-distance flyers, especially seabirds like albatrosses, phalaropes, and shearwaters, which spend most of their lives cruising the oceans, their great backyard. The Sooty Shearwater is special because not only is it a great long-distance flier but it also dives and swims underwater to depths of 150 feet.

The record for the greatest long-distance flight, however, is held not by a seabird but by a shorebird. A satellite-tagged Bar-tailed Godwit named E7 flew across the Pacific Ocean from Alaska to New Zealand nonstop in nine days, a total of 7,145 miles. She closed down one side of her brain and then the other in order to "sleep," and by the time she made it to New Zealand she had used up half her body weight, all the fat reserves she had stored up in the Arctic before taking off.

No creature can top this record for endurance. For a human being, it would be like running more than 43 miles an hour for seven days straight. E7 was not alone in her journey; about seventy thousand Bar-tailed Godwits make this flight twice yearly, but their numbers are declining dramatically. One of the suspected reasons for the decline is that on the return trip to Alaska in the spring the birds stop to refuel on wetlands and mud flats along the shores of the Yellow Sea, between China and the Koreas. Rampant development has been changing the flats, obliterating a major staging ground for the birds. One seawall in South Korea is more than 20 miles long, resulting in more than 154 square miles of lost mud flats. While sediment from the dammed-up Yangtze and Yellow rivers is offsetting some of the food source decline, it is doubtful it will compensate for the traditional flats. In addition, with the fear of rising tides globally, it is unlikely that mitigating development along shorelines will stop. This is just one of the myriad complexities of trying to save species while meeting the needs of mankind.

But back to Hawaii: reaching the islands millions of years ago as the early bird ancestors did was difficult but hardly impossible. Bats came too. The Hawaiian Hoary Bat is the only endemic mammal on the islands; it has been there for a long time, as fossils attest, but it is an endangered species today. Some insects arrived too. Even spiders, not technically fliers, get a lift on the filaments they spin, riding a storm front for hundreds if not thousands of miles. And some butterflies, seemingly so delicate, are formidable long-distance fliers.

These early colonizers of the islands lived in a kind of paradise. There were no predators and no biting insects. Each island had its own riches and birds found their own niches in the abundant flora. With no mammals other than the bat and no reptiles to eat them, the Hawaiian Islands became bird islands. A single ancient finch fathered dozens of different kinds of honeycreepers, an example of radiation in species that puts other island communities in the world to shame. Darwin should have come to Hawaii, not the Galápagos, for variation within species. His understanding of what he called natural selection to explain how the Galápagos finches evolved so individually might have been accelerated had he stumbled onto the fifty-five kinds of honeycreepers in Hawaii first.

All things alive on Hawaii were found nowhere else on earth before people arrived. This endemism is common to islands but extreme on Hawaii because of its remoteness. The honeycreepers probably descended from an ancestor of a Eurasian Rosefinch between 5.8 and 7.2 million years ago, according to the biologist Dr. Heather Lerner at Earlham College in Indiana. The Rosefinches are known to pick up in huge flocks and look for better food sources, especially when there are too many of them, a phenomenon called irruption. They could have flown across the Pacific on storm winds and colonized more than one of the islands.

The Ring of Fire forms an arc that hugs the Pacific coastlines from Chile to California, north to Alaska and the Aleutian Islands, west to Japan and Java, and south to New Guinea and the Pacific islands east of Australia. It is the most active earthquake area in the world, with 90 percent of the earthquakes and 75 percent of the volcanoes on the planet. The subterranean plates are constantly in motion, creating friction and ultimately eruptions of one kind or another. Earthquakes are more unsettling than volcanoes, because you cannot see them, you feel them. Volcanoes thrust into the air with fire and steam, and a kind of majesty.

Ed and I were birding parts of the Big Island in Hawaii for most of the day, catching sight of the Hawaiian Hawk, or Io, in a myrtle tree peering below for insects and little rodents. This small graceful Buteo was one of the first to make the U.S. endangered species list in 1967, when the population was down to a few hundred birds. Its protection was successful and the bird rebounded to several thousand today, engendering a new debate about delisting the species.

It was 1986 and we headed up the long drive to the top of Kilauea, one of the most active volcanoes in the world, spewing continuous steam and magma since 1983. The volcano may be half a million years old. From rustic Volcano House at 1 Crater Rim Drive we watched the roiling mass of fire in the caldera below and trails of molten lava making their way to the sea, where they arrived miles later with a steamy hiss of expiration, the breath of the goddess Pele, whose home this is. After dinner we left word to be called if there was any new action in one of the vents, and at 4 a.m. we were awakened by an enthusiastic operator who said, "The volcano is erupting." We jumped into our clothes and drove through the dark searching for the flue of fire and brimstone. You would think it would be easy, but it took us a half hour, driving aimlessly on unlit roads, before we pinpointed a roar like a freight train barreling down the tracks, and saw the glowing red column of magma against the horizon thrusting from a huge lava field into the predawn light. The column must have been fifty or sixty feet high, and you could feel the heat of the orange flame even though it was a good five miles away. We parked and trekked for several hours across old smooth lava beds and then newer spiky ones that threatened to pierce our ankles if we broke through the crust. The noise drew us closer and closer, but the heat of the air, particles of soot, and empty stomachs finally forced us to turn back before noon. We saw no life on these lava fields created from recent eruptions—no ferns, grasses, or insects. The lava, rich with ash and its nutrients, was just beginning to break down as weather eroded the surface. It would eventually become some of the most fertile soil on earth like other areas of the islands, ready to accept seed.

I have never been comfortable around fire, fearing it will escape the burning logs with a single spark. It must always be watched, like a stranger in the house threatening to steal all.

At fourteen I read about Joan of Arc, also a girl of fourteen, burned at the stake. The cruelty and pain of her death seemed far more horrible than the guillotine or the rope. It frightened me. At sixteen I was determined to play Joan and auditioned for the director Otto Preminger for the movie, a part that ultimately went to Jean Seberg. When I was twenty-five the director Ed Sherin, who later become my husband, offered me the part of Joan of Arc in George Bernard Shaw's *Saint Joan* at Arena Stage in Washington, D.C. Joan's death by fire happens offstage, but the prior trial and Joan's denunciation of her beloved saints was a spiritual agony all its own. In the finale Shaw exquisitely releases the actress and the audience with an epilogue played in heaven. Joan learns she has been canonized and suggests to her dead companions that she now return to earth as a living saint, a prospect they view with alarm.

"O God that madest this beautiful earth, when will it be ready to receive Thy saints? How long, O Lord, how long?" says Joan as the curtain falls. How long, indeed? Joan has never been forgotten; though she is long dead, her spirit still lives. It is that promise I remember—the life after death. The seed of life flowering out of lava ash.

I literally caught fire only once. In rehearsals in Scotland, while attending the University of Edinburgh for my junior year, I pressed myself and the long dusty rehearsal skirt I wore as Ophelia up against the warmth of a gas heater. The whoosh of the flame was instantaneous, engulfing the old skirt in seconds and me in it. My Hamlet raced across the room, embraced me, and threw me to the floor, rolling us over and over again until the skirt was a scraggle of disintegrating black fiber. The tights I wore saved me from burns. The shock of it never left me.

> *Some say the world will end in fire,*
> *Some say in ice.*
> *From what I've tasted of desire*
> *I hold with those who favor fire.*
> *But if I had to perish twice*
> *I think I know enough of hate*
> *To say that for destruction ice*
> *Is also great*
> *And would suffice.*

So said Robert Frost, and I concur: ice will suffice, thank you very much.

———

The road to Hosmer's Grove on the Haleakala Crater from a sun-drenched beach motel an hour below was becoming more and more gloomy. At about five thousand feet mists began sweeping across the slopes and I almost missed a pair of Maui's rare and only native geese, the Nene, combing the grasses for a morning meal of seeds and flowers. This is a small, handsome goose, descended from our ubiquitous Canada Goose, which must have settled down with relief after an endless flight 500,000 years ago. The Nene thrived and speciated with no predation until the Polynesians arrived with their little pigs, dogs, and hitchhiking rats around 500 AD, also with considerable relief after years of paddling their canoes across the South Pacific. Four hundred and fifty years later the Nene population was down to just thirty birds. A protective effort began in 1952 and today the population stands at about 2,500 birds on four islands. They were a welcome sight at dawn on this birding day.

What was not welcome was the thick fog that greeted me when I finally parked at the empty lot of the grove. I had dressed for the chill at 6,800 feet but not the impenetrable wall of gray that enveloped me as I entered the small forest called Hosmer's Grove, a collection of native and non-native trees that Mr. Hosmer planted back in the 1800s for a timber industry that never took off. Birders are confronted with challenges all the time—this was a cloud forest, after all—so I ventured in hoping for a call, if not a song, from one of the honeycreepers. I thought of Bret Harte's "A bird in hand is a certainty. But a bird in the bush may sing."

I actually love fog. It can be eerie, but somehow the more you enter into its realm, the more you receive from it: senses you didn't expect to use, like the touch of it against your skin and the sound around you. It all magnifies.

So I was hoping for the audible life of the birds to come through in the grove. I had listened to tapes identifying some of their calls, but I was not confident in my knowledge. This was before CDs and apps were invented. I was on my own in a new world with birds I had never seen. The large trees cocooned the grove and moisture dripped from needles of pine, fir, and spruce. Then something moved above me in the soft light and barely four feet away were red feathers and then a squeaky song not unlike that of our Red-winged Blackbird back home. The great curved beak made the I'iwi unmistakable. I had my first honeycreeper. In the next hour I had close-up views of the greenish 'Amakihi, and the rosy 'Apapane sucking nectar from scarlet 'ohi'a flowers. The fog allowed me closer than I was ever normally privileged to be while bird watching.

These three honeycreepers are more easily seen than the other fourteen still in existence, all that is left of the original fifty-five. Fully eleven species are on the endangered list and the little Po'ouli of Maui is virtually extinct. Discovered in 1973 by some University of Hawaii students, the chunky black-faced bird has not been seen since 2004.

The Hawaiian Islands are known as the extinction capital of the world. No place on earth today is losing species so fast. Those Polynesian pigs, dogs, and rats did a job on the islands, rooting in the earth, eating most creatures, and devouring eggs. But it was the whalers and stopover sailors in the early 1800s that changed the landscape forever by introducing mosquitoes, specifically a variety called *Culex quinquefasciatus*. There were no biting insects in the paradise that was Hawaii before man came ashore. With limited insect eaters the mosquito proliferated, infesting the lowlands. The birds had no immunity against avian malaria and so those living below five thousand feet succumbed to the disease and began the long road to extinction.

The three honeycreepers I saw in Hosmer's Grove survive because the Culex Mosquito doesn't breed above six thousand feet. However as the climate continues to warm the globe and the mosquito climbs the slopes, the future of honeycreepers, indeed all birds with no immunity, remains in doubt.

There are protective measures under way: the eradication of pools and pig fallows where the mosquitoes breed and the vast spraying of lowlands, which has its own negative impact, particularly on shorebirds like the Hawaiian Black-necked Stilt.

And there is hope that at least one honeycreeper, the ʻAmakihi, is developing some genetic immunity. Whether this can happen in time for the species to remain viable is speculative.

In the beginning when the volcanoes were rising from the ocean floor, creating the islands that became Hawaii, green and gorgeous, the paragon of all earthly delights, native species arrived on average once every thirty thousand years. Today a new species arrives with the thronging crowds about every twenty days. Mostly they are small and of an insect variety, requiring more deadly insecticides to control their expansion.

Today there are hardly any native species of animals, plants, or birds below one thousand feet, due to human incursion. Paradise lost. "Maui" means "God of a thousand tricks" in the sacred lore of Hawaiians. Perhaps he has one up his sleeve for the resurrection of exquisite and doomed creatures.

13

Galápagos

Dragon bodies were everywhere. The volcanic headland was covered with fearsome-looking Marine Iguanas that blended perfectly with the sheen of black rock they rested on. Sally Lightfoot Crabs housed in shells of neon orange wove their way through them to ocean crannies. My granddaughters, in short shorts and tees, stepped gingerly over the spiky lizard tails, snapping pictures of the layabouts, while the huge male bobbed his head up and down and opened his wide mouth, exposing a cavern of pink inside. He looked very scary and seemed about to charge, but he didn't—it was all for show.

The wonder of the Galápagos is the benign nature of the birds and beasts. If there is an Eden on earth, this is it. Despite a few hundred years of sailors stopping by for birds and tortoises, loading up their ships for the next whale hunt, the animals remain sanguine around human beings. Tortoises, of course, do not move quickly; perhaps if they did, they would run like hell.

In 1820, after a Sperm Whale bonanza was reported by the Nantucket whaleship *Globe,* hundreds of whalers put in at the islands to pick up tortoises for food and oil. They flipped them on their backs in the hold, stacking them side by side and on top of each other, and they lasted for months. Had the whaling business persisted, the tortoises would probably be extinct, as the population plummeted from 250,000 to about 20,000 today. The very fact that they are big and can last without food and water for months is what brought them to the islands in the first place. A few million years ago a pregnant South American tortoise rode the Humboldt Current six hundred miles west to one of the eighteen islands. Her size allowed her to keep her long neck extended so she could breathe as she floated along for weeks or months. Her progeny dispersed to other islands, creating subspecies, ten of which still exist today. When my granddaughters and I were there, there was an eleventh species represented by just one individual from the island of Pinta who was called "Lonesome George," then in his dotage at almost one hundred. We watched him inch around his enclosure on Santa Cruz Island, his head stretching out from under his steep saddleback shell as he reached for leaves. He looked like ET with his wizened skin and big wise eyes. Six months later, on June 12, 2012, Lonesome George stopped breathing, ending millions of years of his lineage.

Vita and Isabelle are cousins, both born in 1999, at the close of one century and the promise of a new one. When my grandchildren turn twelve I take them wherever they wish to go in the world. It

Lonesome George near the end, in his enclosure at the Charles Darwin Research Center, Galápagos, 2011

is a rite of passage for me as much as it is for them because I get to spend time watching their young minds absorb the wonders of a new world. The girls deliberated for more than a year, debating the merits of Paris as opposed to Hawaii or Thailand, and finally chose the Galápagos, which made me happy.

In the same year they were born I made my first visit to the islands. As a trustee of the Wildlife Conservation Society, I celebrated the centennial of the Bronx Zoo with my colleagues on the islands. We called it "In the Wake of William Beebe," who was the zoo's first curator of birds and who wrote eloquently about his time in the Galápagos. We visited the islands he researched in the 1920s and where WCS still plays a role.

Beebe was an exceptional man. The natural world fascinated him in all its intricacies. He had no patience with people who were bored and once said, "Boredom is immoral. All a man has to do is *see*. All about us nature puts on the most thrilling adventure stories ever created, but we have to use our eyes." Beebe was one of the first scientists to talk about the relationship of organisms to one another—he was an ecologist studying ecosystems before the terms came into existence. He traveled extensively and wrote books on pheasants in Asia, coral reefs off Haiti, fish in Bermuda, jungle wildlife, and the Galápagos. He feared the loss of species, and one of his most eloquent statements is a prayer for the ages: "The beauty and genius of a work of art may be reconceived though its first material expression be destroyed; a vanished harmony may yet inspire the composer; but when the last individual of a race of living things breathes no more, another heaven and another earth must pass before such a one can be again."

Will Beebe was a celebrity in the 1920s and '30s. He and a friend made thirty dives in their "Bathysphere." The metal globe weighed five thousand pounds, was four feet nine inches in diameter, and had three small portholes of fused quartz. In 1934 they went deeper than anyone had ever been, 3,028 feet below the surface of the ocean. Beebe compared the luminous fish he saw on those dives to stars in the night, and said that exploring the ocean depths was like exploring space. Seventy years later we still know as little about the oceans as we do about outer space. The oceanographer Sylvia Earle, who was influenced by Beebe, says 95 percent of our oceans are unknown to us.

So, in the wake of William Beebe those of us from WCS set off to follow his Galápagos Islands journey: from San Cristobal to Genovesa to Fernandina and Isabela. Then on to Santiago, Bartolomé, Santa Cruz, and South Plaza. Bottlenosed Dolphins rolled in the waters, shearwaters skimmed the air above the waves, and silver fish skipped the surface.

A few children and grandchildren of fellow trustees were on that first trip. My husband and I noted how special the islands were for them. The tidal pools allowed close observation of crabs, small fish, and baby octopus while the rocks were covered with herons, oystercatchers, Yellow Warblers, and lazing sea lions and iguanas.

No need to fly from predators, so the Flightless Cormorant evolved in the Galápagos.

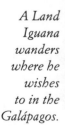

A Land Iguana wanders where he wishes to in the Galápagos.

Everything was right out in the open, with no bars as in a zoo. We walked paths that brought us next to nesting Blue-footed and Red-footed Boobies, and scrub forests of Frigatebirds with puffy chicks. We watched Flightless Cormorants, found only in the Galápagos, preening on the rocks. And we swam with sea lions. There is a primal contentment in being close to animals, a sense that this is the way it is supposed to be, the way described in Genesis.

Twelve years later, Isabelle, Vita, and I boarded National Geographic's ship *Endeavor* in Guayaquil, Ecuador, a few days after Christmas 2011. This was a voyage for families, and a third of the passengers were children between the ages of four and fifteen. At least a dozen pubescent boys and girls from different parts of the United States, Ireland, and Australia raced each other around the decks, cannonballed into the pool, and hung out in the lounge playing games after dinner. We had come to see the animals, and for my girls the young human males were the most interesting of them all—well, that's another story in Genesis.

Galápagos was never truly settled until well into the twentieth century, unlike Hawaii, which the Polynesians may have visited as early as 800 BC. Both archipelagoes are volcanic, but Hawaii's earliest islands are twenty-eight million years old while Galápagos's are three million. There is ongoing volcanic activity on both. Hawaii sits in warm waters, while Galápagos, near the equator, is in the path of the cold Humboldt Current traveling north from the tip of Chile. The upwelling created by the Antarctic water colliding with warmer surface water makes the Humboldt the richest ecosystem for marine life in the world. Small fish like anchovy, mackerel, and herring feed on nutrients brought to the surface, and large fish like tuna feed on them. These nutrients are particularly abundant from May through November, resulting in a profusion and diversity of marine life and birds. It is the northernmost outpost of the penguin.

Millions of years of evolution, of adaptation to life without significant predation, and of abundant food sources, have made the animals unafraid. Some of them are even bold. On Floreana it was hard to bring my water bottle to my mouth without an endemic Floreana Mockingbird jumping on top for a sip.

This is the same Mockingbird that Charles Darwin collected on his *Beagle* voyage in 1835, one from each island he visited. It was the small differences in the Mockingbirds that he wrote down, not the Finches, which he thought were unrelated to each other. He didn't think the finches were adaptive variations of the same finch ancestor, so he collected them without bothering to note which island they came from. The idea of species evolution didn't begin on that voyage of the *Beagle*. He had clues when he was in the Galápagos; he even wrote in his journal, "The different islands to a considerable extent are inhabited by a different set of beings . . . my attention was first called to this fact by the Vice-Governor, Mr. Lawson, declaring that the tortoises differed from the different islands, and that he could with certainty tell from which island any one was brought." This didn't fully register

One of Darwin's Finches,
the Cactus Finch

at the time with the young Darwin. It percolated but didn't become a real insight until he was back home in England, where the lab dissected his specimens and he learned that the thirteen finches were descendants of one ancestral species, pure examples of adaptive radiation because they filled different ecological niches on the islands. He mentions the finches in his *Voyage of the Beagle* in 1839 but, curiously, not in *On the Origin of Species,* which he finally published in 1859, twenty-three years after his *Beagle* trip ended. He and Alfred Russel Wallace, in a synchronicity common to great ideas, and a desire to be first in print, published parallel theories at the same time.

The first jump into the water was a jolt despite our half wet suits. Vita's long slim body shivered as she snorkeled next to me. A huge Green Turtle glided by and then a Spotted Ray, like a great bird in slow motion. There were clownfish, groupers, and Tiger Eels. Then a Galápagos Penguin shot by like a bullet. A six-foot Marine Iguana chomped on algae, tearing it from a clump on a submerged rock face. Life underwater was bucolic until a mammoth sea lion butted my side as he streaked by. This was a warning from the "beach master" protecting his harem that I was in his territory. The Galápagos does have its darker side, not only human incursion but predation by owls and the Galápagos Hawk of small lizards, bird's eggs, and insects. The owls and hawks have little fear of humans. A Barn Owl slept blissfully three feet above us as we changed into our bathing suits, and Darwin, collecting species, said of the Galápagos Hawk: "A gun is here almost superfluous; for with the muzzle I pushed a hawk out of the branch of a tree." Underwater, it is catch as catch can, as it is anywhere in the great ocean, big fish after little fish.

Vita and Isabelle were scared of sharks. I was too. How could one not be after seeing Steven Spielberg's movie *Jaws*? Life for me is marked "BJ" and "AJ"—before *Jaws* and after *Jaws*. As a girl I swam unafraid in the waters of Surfside, Nantucket, in front of our summer cottage. A fin in the water was not an uncommon sight—we knew it was probably a harmless Nurse Shark feeding at the bottom on calm days, or if the sea was rough and a dark fin appeared, we would just sit it out until the Blue Shark or Mako or whatever it was passed by after the school of Bluefish it was chasing. Swimming in the moonlight, following the path of sparkles it made on its watery beam, was one of my greatest joys. On dark nights the phosphorescence glowed and lit our legs beneath the waves. I knew kids who died from polio during those years in the 1950s, but I never knew one to die from a shark attack.

In the 1970s Ed and I watched our own children swimming in

the surf as carefree as I had been at their young age. Then in 1975 we saw *Jaws*. The scene where the girl is swimming at night and is pulled under by the Great White Shark made a searing impression on me, one I have never been able to erase despite telling myself it's just a movie. How can people say images do not make a difference to lives? Hasn't the advertising business known this forever? My courage failed me. I never dove into that watery path of moonlight again.

So it was with some trepidation that I joined others in a Zodiac boat to swim with sharks. Isabelle and Vita chose to stay on the ship; I didn't blame them. We motored to a deep channel between a one-hundred-foot pinnacle and Kicker Rock. Our guide was a beautiful young Ecuadorian woman who had no fear, but she also had no wet suit or snorkel, having left them both on the ship. *She* wouldn't go in, but she kept urging *us* into the water. No one moved, everyone busy fiddling with his gear. I was the oldest and figured I had the least to lose and so found myself slipping off the tubing into the cold unknown. When I put my face mask into the water I was stunned. I would have gasped but for the breathing tube in my mouth. Below me, down to my right, to my left, and all around were more sharks than I could count. They were larger than I was, and streaming by continually. I was mesmerized, quite unable to move for a few minutes. They were so close, but they weren't paying any attention to me; perhaps they had eaten well and were just out for a cruise. I boldly took a deep breath and closed my snorkel intake, swimming ten feet down. Galápagos and Whitetip Sharks passed by with no acknowledgment; the Blacktip Sharks were still a few feet lower, and even lower were some Hammerheads.

Once in Veracruz, Mexico, I had walked way out on the shallow sandy bottom of the great bay looking for some cool relief from the heat. My brother and a friend sat on the sand getting smaller and smaller as I searched for the deeper cool. My head and shoulders were just above water when a large Hammerhead loomed into view. I stopped, barely breathing, and didn't move a muscle. The eight-foot body circled me leisurely five or six times, three feet away. Then the strange shark with eyes at the outer tips of the "hammerhead" decided I was not worth her time and moved off with hardly a swish of her tail. I still had a residual fear of Hammerheads.

My diving "buddy," a young woman whom I had just met in the Zodiac and who had followed me into the water, suddenly dove, paddling her flippers thirty feet down. She reached out a hand and touched the top of a Hammerhead. I was shocked. The shark did nothing, and my buddy rose to the surface with a big smile. "I love Hammerheads," she said.

The fear of sharks that had been with me for decades simply left, and fascination took its place. They were beautiful. Whitetips have a terrible reputation as vicious killers, as do Great White Sharks. I know they have killed people. They are the top predators of the ocean just as the great cats are of the jungles. But my experience was benign. They patrolled the deep water but didn't seem on the hunt—they just glided by, as did Green Turtles and Spotted Rays. My companions and I swam with these sleek creatures from one end of the deep channel to the other where the Zodiac waited. In the end it was a swarm of minute jellyfish, their tiny filaments stinging like mosquitoes, that drove us from the water, not sharks.

This was a profound experience for me. I felt liberated. I had hauled the burden of fear around with me for so long that it was part of me, and now it was gone. I felt like a kid who jumps off a high diving board for the first time. In the Galápagos Islands, where animals have no fear, I too gave it up. I began learning all I could about sharks.

One of the films I saw was *Sharkwater* by a young Canadian diver and filmmaker named Rob Stewart. Sharks are maligned creatures, accused of being primitive because they are four hundred million years old, when in fact they are highly sensory. They have two more senses than human beings: a lateral line along their bodies like all fish, which tells them what is moving in the environment around them, and an ability to detect the electromagnetic fields of all matter. These senses, and being able to move through water with no friction, make sharks apex predators of the oceans. They keep the ecosystem in balance. Because sharks kill so effectively and sometimes we are their inadvertent prey, we fear them and hate them. It makes sense, but in retaliation we have killed millions and millions of them, to the point where more than a third of them are listed as endangered by the International Union for Conservation of Nature. Ninety percent of all Great White Sharks are gone and the Mako and Porbeagle are in such trouble that all three are on the IUCN's Red List. They mature late and have few pups. The Mako female breeds when she is nineteen to twenty-one years old and has between 4 and 18 pups; the Porbeagle female is mature at six to eleven years and whelps only 4 pups a year. The Blue Shark, by comparison, can have as many as 135 pups in a litter, and females are mature before they're five years old.

Sharks have died at the hands of sports fishermen and as bycatch of commercial long-liners, but mostly they die from the global business of shark finning. The shark population has been decimated to feed the Asian market for shark fin soup. *Sharkwater* exposes this deadly practice in countries like Costa Rica, where it has been outlawed but continues illegally, and in the Galápagos, the most protected marine reserve in the world, where illegal long-liners elude the few patrol boats. Even in paradise the devil finds a way.

The good news is that every day more countries are outlawing the sale of shark fins, Asian celebrities are exposing the needless killing of sharks, and chefs are finding substitutes to satisfy the Asian appetite. The irony is that shark fin has no real taste and is put into a chicken broth to give it flavor.

I watched my sweet twelve-year-old granddaughters as they absorbed the wild and wonderful islands we visited. Their favorite afternoon was spent bodysurfing the waves with their new teenage friends and the occasional Sea Lion, while I watched from shore, as intent on a rare Lava Gull at the tide line looking for food as I was on them. Human and animal pursuits are virtually the same. We eat, we sleep, we seek shelter and companionship. We court, we have families, and we play. The baby Sea Lions were playing on the backs of their mothers, while my girls were playing on the backs of waves.

My granddaughters Vita and Isabelle after swimming with
a sea lion in the cold ocean, Galápagos, 2011

On the last day we walked through a misty drizzle in grassy fields where Galápagos Giant Tortoises roamed. Their domed black shells glistened in the rain as some of them retreated inside. Others seemed curious and gazed on us with eyes that had seen much in a hundred years or more: capture, starvation, drought, and plenty. The girls were only twelve. What would the world be like as the twenty-first century rounded out *their* old age? Some of these Tortoises might still be alive. The fact that they had already survived so much gave me hope—that and my granddaughters' commitment to keeping them so.

Galápagos became a national park in 1959 to mark the centennial of Darwin's book. There are about thirty thousand people living on a few designated islands and the population is growing. Tourists come by the hundreds of thousands now but are regulated to a limited number at any given time on boats and paths to keep the islands free of disease. Still, people bring microbes on their footwear, plant spores on their headgear, and any manner of tiny new flora and fauna to the islands. It will be a battle to keep it pristine in the future. The feral goats, pigs, and cats are being brought under control through eradication so that the endemic species will continue to thrive. This is not without controversy; some people do not condone the killing of any animal for the sake of another. However, the Galápagos is a highly managed environment today. Two hundred and fifty thousand invasive goats were killed by sharpshooters on Isabela and other main islands in order to allow the vegetation and water pools needed by the endangered Tortoises to flourish again. By 2006 they were declared free of large introduced mammals. These are hard choices, but without such stringent control Galápagos would lose its native species as Hawaii has and continues to do as climate change drives them toward extinction.

In 2012, helicopters dropped poison on the island of Pinzón, killing the invasive rats that had wiped out most of the rare Tortoises there. In 2015 ten hatchling Tortoises were found on the island, the first in a hundred years, promising survival of the species, reversing Darwin's trajectory of "survival of the fittest," that were the indomitable rats we humans brought with us.

14

Newfoundland

We were bound for Francois, five other passengers inside fighting waves of nausea as the little ferryboat heaved on the black water, hugging the coastal cliffs of Newfoundland. I was never a victim of seasickness, but I sat outside on top of one of the monstrous cargo containers and let the wind blast my cheeks. My legs were hanging down in the space between the metal boxes before an adjacent one began its frightening slide toward me in the swell. I jumped down just as metal slammed against metal, and made my cautious way on the wet deck to the hatch and the reeking stench inside.

The ride from Burgeo to Francois takes four hours. It is the only way to get there, and has been for centuries. There are no roads through the dense underbrush and wetlands in the heart of Newfoundland. At one time there were perhaps thirty outports on this south coast, the tiny villages of those who made their livelihood from the great fishing banks of the North Atlantic. These are the folks who hauled in hundred-pound Cod on their lines and Halibut weighing four times that. They filleted and salted the fish in these outports and sent them abroad for the table of French kings and the streets of Philadelphia. The North Atlantic is an unforgiving ocean, and these are some of the hardiest people on earth.

The lure of the Grand Banks, one of the most productive fishing areas on the planet, has been irresistible since the 1400s when the

Portuguese and Basque braved the crossing from Europe in their small boats. Cod, Halibut, Haddock, Herring, Mackerel, and Red-fish (Ocean Perch) all thrive on nutrients created by the cold Labrador Current from the north and the warm Gulf Stream from the south. Capelin, small baitfish, feed the larger fish and the millions of seabirds that congregate on rocky cliffs to breed.

We stopped at one of the few remaining outports, Grey River, to deliver goods as my fellow travelers gratefully disembarked. I took the opportunity to talk with some of the women gathered on the wharf in the late afternoon sun while the cargo was hoisted over. The ferry comes only a few times a week and the slip is a gathering spot, the only game in town for some of the 130 residents who live here. The teenage girls were rather chic in their tight black jeans and leather jackets, with studs in their noses, cuffs on their ears, and purple hair just like any counterpart in Brooklyn; like teenagers anywhere, they were not terribly interested in talking to me. The older women responded curtly to questions about life in the village. "We don't want for anything," one said. "Had the Internet for seventeen years." It was 2010. They had been wired since 1993! That's longer than 90 percent of New Yorkers. Because of the remoteness of the outports the Canadian government takes care of them with special services, including helicopters in case of emergency.

It was another hour or two before we came to the long fjord heading into Francois. The evening sky was hung with gray clouds dappled by a lowering sun as we motored between two steep hillsides. A tiny lighthouse perched above to direct errant sailors. The fairy-tale town finally came into view nestled beneath a sheer granite wall a thousand feet high that curved to embrace fifty or sixty gaily painted houses. They were stacked like dollhouses one above another on the scree, climbing to the stone face of the wall. The sun took a last spill over the pinks and blues, the reds, oranges, and purples of the houses and disappeared behind the mountain.

My host for the night, Ethel Andrews, met me with a smile, grabbed my bag, and led me on a narrow winding boardwalk uphill toward her aqua home. In a tiny kitchen she served her husband,

John, and me a comforting meal of sandwiched beef smothered in gravy, mashed potatoes, carrots, and peas. Then John headed back down to his fishing shack to hook his lines for the next morning. He would place hundreds of baited hooks that day and in the hours before bed. He lost a little finger to blood poisoning while placing a hook once, and the fingers on his right hand are bent from Dupuytren's contracture. Still he baits faster than anyone I have ever seen.

Ethel took me for a walk up the mountain to the community's graveyard, a source of pride and comfort to the 115 who live in the village and know everyone who has died and all who are born. A slim waterfall spilled from a lake above the granite wall, making a shallow pond in the heather and grasses. A few ducks circled about and one or two frogs croaked in the reeds. The June sky was light until 11 p.m., but we retired early, as John would be up and in his dory before 4:30 to reach the fishing grounds at dawn.

John, at sixty-three, has been fishing all his life. He is one of only five fishermen left in Francois (pronounced "Fran-sway," Ethel tells me, the legacy of England's domination of the Maritimes and their disdain for the French). The collapse of the Atlantic Cod in 1992 led to a moratorium on taking the fish by 1995, and the industry never recovered. Neither did the Cod, not as before. The catches are low, and the giants that used to be pulled from the sea and could feed an army are gone forever. I treasure an old postcard found in shops from Cape Cod to Newfoundland, of a nine- or ten-year-old boy holding by the gills a gigantic Codfish a good foot taller than he. Halibut, the largest flatfish in the world, has suffered the same fate and is now on the endangered list. John was excited a few years ago to catch a 150-pounder within his quota, but it fetched only $400 at market, a pittance compared to the price the fillets of a top restaurant might command in New York.

It is not possible to make a good living as an independent fisherman anymore; the huge trawlers and the global corporatization of the seas have brought an end to the ocean lives of these fishermen as they have to the Cod. Like the Passenger Pigeon, the most plentiful bird in the skies before it was shot to extinction in the early part of the twentieth century, the Cod produced bountifully in the cold waters of the North Atlantic and could have continued to feed the world's growing population had it not been for the marriage of technology and factory ships that overwhelmed the traditional fishing fleets. As Mark Kurlansky points out in his book *Cod,* it was the unending abundance of the Codfish, the belief that it would go on flourishing forever, that did it in. The trawlers with their huge nets dragging the ocean waters, equipped with devices to ferret out the fish wherever it might lie, including spawning areas, continued to exploit the Cod, never believing it would crash, until the government finally called a halt. The United States closed Georges Bank in 1994, the richest grounds in New England and Nova Scotia, and the Canadian Grand Bank soon followed. There was outrage from the fishing industry, which said it would go under, jobs would be lost, and there would be no fish in the markets. Independent fishermen quietly cheered the decision, knowing firsthand, as did John Andrews, the toll taken by the huge trawlers and corporate entities. But the moratorium was just that: an end. It was an end to the great Cod fisheries of the North Atlantic, an end to the way of life of Newfoundland outporters, and an end to life under the ocean where Cod had been an apex predator.

Human beings are nothing if not adaptable, and during the moratorium there were other fishing grounds with other kinds of fish, such as Pollack, to be harvested, and there was the nascent business of farming wild fish. Aquaculture took off. Today as I write this, fully 50 percent of all fish consumed in the world is farmed, from family ponds harvesting Tilapia in Malaysia, to man-made trout ponds in Norway, to netted pens of salmon off the coast of Chile. I would not have believed that the migratory Cod, needing deep cold oceans while adult and warmer waters for spawning and young, could ever be farmed. But I was wrong.

Newfoundland had Cod ranches in the ocean in the early 1990s. Brian Johnson taught three hundred fishermen from fifty-two communities devastated after the fisheries collapsed how to farm Cod in pens. They raised them to ten pounds before selling them, and Brian believes the business would have thrived if not for the controversy that ensued over aquaculture. He still believes it can be done right and is the future of fisheries. Now Brian manages the less controversial Scotian Halibut, an inland nursery, in Nova Scotia.

If we are going to feed a global population of ten or eleven billion by 2050 it will be necessary to farm the protein-rich life of the oceans, just as our ancestors domesticated land animals for consumption. In Paul Greenberg's book *Four Fish* he discusses the four mammals—sheep, cows, pigs, and goats—that we husbanded for our use. These animals, along with agriculture, ended the practices of hunting and gathering in most of North America and Europe. If you travel south in the Western Hemisphere, Alpaca and Guinea Pig are the domesticated staples of the Peruvian diet. Hunting and gathering still goes on in many areas of the world, but as mammals and birds become scarce and productive habitats are obliterated because of climate disruption, people will seek other means of survival. They will turn to the oceans, the only vast unknown expanse left on earth.

Greenberg suggests that four fish will be staples of our diet just as beef, chicken, and pork are: salmon, sea bass, cod, and tuna (not the Bluefin species). I am not so confident that cod, even Pacific Cod, and Yellowfin Tuna will make it to the next century. With four fish as diet staples, is this the death knell for the exquisite diversity of taste we have had with so many different kinds of fish? Probably. Leave it to chefs to create culinary variety out of relatively few items. But I will miss the remarkable buttery flavor and texture of Copper River Salmon when it is rushed from the Alaskan rivers to the kitchens of the Pacific Northwest every May. No farmed salmon I have ever tasted can compare. However I put nothing past human ingenuity.

Ethel and I were there on the dock when John returned with his catch before his lunchtime at noon. The other dories had already docked and their catch had been weighed and logged by the government worker from the Department of Fisheries and Oceans, or DFO. John had a dozen twenty-inch Cod, a few ocean catfish, and some colorful redfish—half the haul he had hoped for. When he heard an earlier boat brought in two thirty-inch Halibut, his ears pricked up. It is his favorite ocean fish to eat, as it is mine.

I wandered the boardwalks of the town after lunch, chatting with people sitting out in the welcome seventy-degree weather. It was hard sometimes to understand their heavy accent; they dropped their *h*s and added them before *e*s so that "even here" became "heven 'ere." Unlike the dour folk on the Grey River wharf, the people of Francois seemed cheery—all of them, including the children. Little boys skipped along, asking where I came from and where I was going. The village has few young people. The entire school had eighteen students in 2010, and there would be only sixteen the next year. The high school graduated just one boy and one girl, and they had left for the big towns of Corner Brook and St. John's where they could get a job. Still they come back because, as one old captain told me, "Everyone loves Francois in their own way."

The difference in spirit between captive communities in places like these outports is something I had been told to observe by a seasoned traveler long ago. I find it to be true. Just as nations have their personalities, so do these villages. Francois was reputed to be one of the happiest outports—"Always that way," said one person—while another outport fifty miles away was said to be mean: "They kick their dogs."

How this came about is a mystery. Was it one person, a leader in the community long ago, who set the tone, and is this a trait that passes from generation to generation culturally or cellularly? It would be hard to know.

I watched Ethel as she negotiated her computer as fast as a techie at the Apple store, although she is self-taught—there is no Genius Bar in Francois. She checked her Facebook page and Skyped with her daughter and grandchild in Corner Brook, ordered a mop online, and downloaded some books onto her Kindle. She rose at eight or later the mornings I was there because the sun didn't slide down the granite wall until then. In winter there may be only a few daylight hours, and bed is even cozier than usual. They have another home in the next fjord, as many of the outporters do, a place to get away for the weekends where they collect wild mussels in the waters and steam lobster on seaweed in fire pits. They have sing-alongs on the beach during the long summer nights and dances in the community hall during the short winter evenings. All the women knit and quilt.

Ethel's nephew Darren took me out on his boat to see the Newfoundland coastline farther east. Cliffs towered above as a Bald Eagle eyed us from a windswept spruce snag. I was startled to hear the sweet song of a Warbling Vireo from a clump of green in the rugged landscape, perhaps blown off course a hundred miles east, or dispersing due to warmer climate. If he finds a mate with his warble maybe he'll colonize the species for Newfoundland. We motored down a stunning fjord with a sheer rock wall rising from the shoreline to the east. Rock climbers from Britain to California find their way here; you can't keep a subculture down. But there is no infrastructure for tourists, only a home or two where one can stay the night as I did at Ethel and John's—that is, if you are lucky enough to be given the phone number. A few climbers, backpackers, and hunters make it here, but not birders, although this is one of the rare spots where you can see (and hunt) both Willow and Rock Ptarmigan.

Darren showed me a spit with an old try pot on it where whalers 150 years ago rendered oil from blubber. Whaling was once the fifth-largest industry in the world and the United States ruled the seas, with more than six hundred ships carrying smarter harpoons and more precise charts than others. But with the discovery of petroleum, the overfishing of whales, and the increased costs of going to sea, investors turned inland. The entire industry collapsed in a matter of decades. Whales got a partial reprieve (some are still killed for meat), although they have never regained the numbers they had before the eighteenth century. The parallels today are striking. Drilling for oil and gas is hurting environments on both land and sea. The new energies from solar, wind, and wave power are being developed at a rapid pace and could replace fossil fuels in a matter of decades, reducing carbon in the atmosphere and giving nature a reprieve.

An abandoned village stood on the prettiest slope of grassland around, the burnished reds of three dilapidated houses glowing in the late light of day. It was one of the outports that had been resettled, the villagers having been paid by the government to leave decades before. Resettlement was a controversial program that moved more than thirty thousand people from three hundred remote coastal communities between 1954 and 1975. The government of Newfoundland wanted to consolidate services such as education and health care in "growth centers" and encouraged families to move by paying them more than their annual income, which by the 1970s was about $600. All but about seven of the outports on the south coast of Newfoundland were resettled, but Francois, Grey River, Rose Blanche, Grand Bruit, and a few others could not muster the requisite number of people in favor of moving, so they held on. It is only a matter of time before they must capitulate and move as well; with no employment, limited services, and no youth to carry on, the outports are doomed. This is something Ethel and John understand well. They hope to live out their lives in the community they love and in the landscape of their hearts, and that will be it.

Darren and I headed back toward Francois, passing rock outcroppings with seabirds preening on them, settling down for the evening: Murres, Guillemots, Kittiwakes—birds that spend most of their lives at sea, braving the vicissitudes of the wild North Atlantic and hanging on because it is still wild, the province of the invincible. There was a time when the hardiest of human beings braved it in the same way, fishermen like John going out at dawn in their precarious open boats, throwing out lines for the fish they know better than anyone on earth. John still loves these mornings alone, one man in the wilderness of the vast ocean.

Darren gunned the gas throttle of the motorboat, and a cold wake of spray brushed my arm as we raced toward Francois. He tied up at the dock and I scurried up the hill to my last meal with my new friends. As I sat down at the kitchen table, John told me that Grand Bruit, seventy miles to the west, had recently decided on resettlement after many years of division among its thirty citizens. The government offered them the richest deal yet: a single person received

$80,000 for his home, a couple $90,000, and a family with children got a whopping $100,000 to relocate. The final ferry was picking up the last of the people the next day, June 30, 2010. He bowed his head, murmuring thanks to God, then winked at me conspiratorially as Ethel placed our favorite dish, baked Halibut, on the table.

I mourned the loss of these outports on the ferry ride back to Burgeo—the exquisite beauty of the landscape and endless sea, the comfort and grace of living in such tight-knit communities, and the courageousness of the people. This way of life was gone forever, as extinct as the Great Auk, which was last seen in 1844 in these same waters, clubbed to death by sailors. As the ferry pulled into Burgeo, I saw a small group of people standing together on the wharf near a pile of luggage and furniture. Some of them were crying, their arms around each other. "The folks from Grand Bruit," said a fellow passenger.

15

Amazonia

From my diary:

October 2, 2001

A huge moon glowed off the plane's wing as below a cloud cover put the jungle to bed. I am struck just how far this is from the tumult of the last three weeks. The physical distance catches up to us by satellite—CNN is here in my Tropical Hotel room. But mentally the Amazon is another heaven and earth away. I caught an hour and a half of sleep and then . . . took a riverboat tour down the Rio Negro's tannin filled waters to the meeting of the muddy Amazon, a striking contrast . . . We stopped twice along the river, the people's road. Although the same tourist items are sold everywhere they are still using too many feathers (Scarlet and Hyacinth Macaw) and snakeskins and heads.

A boy of 10 held a small sloth, while squirrel monkeys scampered about, one baby the beloved pet of a three-year-old fellow. Three sc. Macaws, 1 hyacinth, 2 caiman, a boa, an anaconda were held for photo ops. I am always torn about this—yes, it is better for the animals to be alive for tourists but will they simply get more and more for photo ops? . . .

One old man in a tourist tent held my hand and said he held Americanos in "regardo"—they know the whole story—they all have satellite antennas on their shacks, all 10 of the river huts.

It had been twenty-three days since the planes crashed into the Twin Towers, killing thousands. The old man gently held on to my hand, his eyes full. It was so startling to me that in this small Amazon village the news was as present as in New York and that this wizened old man let me know he cared, that my eyes too filled with tears. I was profoundly grateful for his sympathy.

This world is smaller and coming together through technology and an understanding that we all affect each other. We affect each other emotionally because human beings are deeply the same. We seek food, shelter, and community—and we play. Everywhere children are playing, even when the situation is desperately poor or war-torn. Pull out a game, a ball, or a book and they join in. In this little village on the banks of the greatest river in the world the children smiled and laughter spilled out when the baby monkey jumped from one tousled head to another. The laughter of children is a balm like no other.

The great Amazon forests affect the weather of the entire earth. When huge swaths are cut, less moisture is held, more carbon is released, and weather patterns change. The biologist Tom Lovejoy was concerned with forest fragmentation back in the 1970s. The Brazilian government was subsidizing the settlement of ranchers and agriculture north of Manaus but ruled that 50 percent of the land had to be kept wild. Tom convinced them to let scientists decide which tracts to cut and which to leave standing. He and his staff and countless scientists have been monitoring the biodiversity of the tracts ever since, making it one of the longest-running studies of its kind.

Tom invited me to come to Camp 41, where scientists stayed and visitors witnessed their work in the forest. He was a conservation biologist at the Smithsonian at the time.

Tom played a pivotal role in my early days in Washington, D.C., as NEA chairman. I met him in 1993 at a dinner party given in my honor to introduce me to the movers and shakers of the town. I sat on a silly stool beside a low coffee table while four media and political giants peppered me with questions. "How do you define art?" began David Gergen. I burbled a boring and theoretical response that initiated more pummeling from the group before they rose as one and left me—a pallid, deflated version of myself, reduced to jelly on the floor. In a corner a few feet away sat a fine-looking bow-tied man with an amused smile, taking it all in. He rose and whispered in my ear, "Welcome to Washington. Let me take you to lunch next week."

I was grateful to Tom for showing me the ropes and becoming my friend. He was immensely successful in negotiating the political scene for science and managed to be a top biologist and ecologist at the same time. Tom coined the phrase "biological diversity," or "biodiversity," back in 1980. He knew that all living things depended on other living things for their existence. The tracts, or reserves, around Camp 41 were revealing just how impacted flora and fauna were by fragmentation of the landscape.

Camp 41 was beautiful, deep in the forest. It was quiet, about eighty-five degrees, and there were no mosquitoes. I walked down to a stream that had been dammed with a few rocks to create a small pool. There I stripped and lay on my back in the shallow water, looking above at the canopy in the late afternoon sunlight. A five-inch Blue Morpho Butterfly wafted above me like something out of *Fantasia,* the sky blue of its wings glowing against the backdrop of green. Frogs and insects kept up a continual chatter while birds flitted by on their way home. I was in heaven.

A dozen hammocks were strung in a line over concrete under the roof of a wooden porch. I put my backpack on one and joined Tom and a handful of other scientists for drinks in a circle of chairs. And then dinner—a most memorable dinner cooked by a local couple over a wood fire. The fish, called Tambaqui, is the most delicious fish I have ever tasted. It is a large Amazonian river fish that eats palm

The quintessential ecologist Tom Lovejoy, in the Amazon in 1989

nuts when the river rises and floods the trees. It has molars for teeth, so it can chomp on the nuts. It was slowly grilled about three feet from the wood fire, infusing its soft flesh with palm oil.

As we rolled into our hammocks Tom said we would be awakened predawn by the call of a Motmot, as reliable as any alarm clock. The entire night was alive with an infinite variety of insect, frog, and monkey music. First light barely made its dim way through the trees when the Motmot softly whooped its wake-up call in stanzas of two, just as Tom had said it would.

We spent the day walking the forest paths learning about the intricate systems of ant species from one biologist and the interplay of Red Howler Monkeys from another. I failed to see the deadliest viper of the Amazon when a young Bushmaster crossed the path in front of me, its dark back blending perfectly into the leaf litter. Tom behind me almost put his boot on top of it. The snake was more alarmed than we were; it took immediate refuge at the base of a tree, coiled to strike should we come any closer.

More than thirty years of studying the isolated reserves of forest within the scrubland of ranching and agriculture has proved that the species count eventually drops dramatically—that fragmented parcels cannot continue to sustain the biodiversity found within large pristine areas. Elizabeth Kolbert writes about ants and the antbirds that depend on them for sustenance in her seminal work *The Sixth Extinction*. She too visited Lovejoy's reserves and followed the expert Mario Cohn-Haft as they looked for marching ants. Antbirds were also looking. When no ants were found Cohn-Haft thought the ants must be in the stage where they stay in one spot for weeks to raise young. In fragmented areas there are simply not enough colonies of ants to sustain the antbirds, so the birds soon disappear.

"When you find one thing that depends on something else that, in turn, depends on something else, the whole series of interactions depends on constancy," said Cohn-Haft. "Things fall apart, the center cannot hold;/Mere anarchy is loosed upon the world," wrote W. B. Yeats.

In an effort to stem the anarchy Tom Lovejoy invited politicians, celebrities, journalists, and anyone else who might be influential to come and visit Camp 41 to experience the Amazon rainforest for themselves. Senator Alan Simpson of Wyoming, a staunch supporter of the NEA in its struggles with his fellow Republicans, came, as did Tom Cruise and dozens of others. Tom Lovejoy has been working tirelessly to effect change at the top, in the halls of Congress, foundations, and international agencies. Without legislation to protect ecosystems they will all fall apart in a "Tsunami of Extinction," as he wrote in *Scientific American* in 2012.

The water comes alive at night in the Amazon. In the bend of hundreds of tributaries of the great Rio Negro, animals collect to feed in the dark. Our canoe glided into deep pools through air thick with insects. The fifteen-horsepower motor purred to a stop and we turned on our flashlights. There before us as far as darkness allowed were the glowing red eyes of hundreds of caimans waiting for a meal, snouts nestled silently below the surface. They had no interest in us; they waited for the Pacu, or for one of the many species of piranha, or for the juveniles of the Pirarucu, the largest freshwater fish in the world, which at maturity weighs up to 440 pounds and reaches 15 feet long.

Their eyes reflected the black water like red marbles on a glass tabletop. These were the Spectacled Caiman, a smaller relative of the American Alligator, plentiful throughout Central and South America, its skin being unsuitable for leather products. The larger, aggressive Black Caiman, however, is hunted for its highly prized skin, despite being an endangered species.

The canoe sliced the water as we returned to our floating lodge displacing fish right and left, many flying through the air into the boat, slapping us in the face on the way. Ah, breakfast! I longed again for the taste of these Amazon fish, the Tambaqui and others that feed on fruit and nuts falling into the water from branches above. They hang out below a nut tree, mouths agape above the waterline, waiting for the seed to drop. The Amazon Basin has more than a third of all the freshwater species of fish in the world, more than three thousand and still counting.

Tom and I had left his Camp 41 and were staying with colleagues from WCS at the Mamirauá Sustainable Development Reserve in the heart of the flooded forest of Amazonia in northwestern Brazil.

As a member of the Conservation Committee of WCS I had come to observe the work of the Brazilian biologist José Márcio Ayres, who was dedicated to the preservation of wildlife in this part of the Amazon. Every year in the rainy season between November and June the forest, or várzea, floods as high as thirty feet. The fish move up tributaries to spawn, their predators following, while mammals scramble to higher ground and treetops. Most of these mammals are well adapted to arboreal living, squirrels, howler monkeys, and sloths among them.

The sloth, the slowest of all mammals, traveling on average six feet per minute, spends its life hanging from branches, sleeping, eating, and even mating upside down. It comes down from its tree only to urinate and defecate once a week, always in the same spot. We watched patiently as one took fifteen minutes to make the trip down, do its business, and make the equally sluggish climb back up. With only leaves for its food, the sloth has limited energy and muscle mass.

The Boto, or Pink Dolphin, patrols the Amazon waterways with grace and speed, looking like a giant blob of bubblegum when it surfaces. The first time I saw one I was on a ship from Rio to Manaus as one of a troupe of actors and singers from the United States and Canada who performed for the three hundred people on the cruise. Patricia Neal, Julie Wilson, Brian Bedford, and a young Anna Bergman were among those who sang or spoke or acted as we steamed upriver; they became friends for life. The two-week venture put in at the old colonial coastal cities of Recife, Salvador, and Fortaleza before entering the huge delta of the Amazon and heading into the heart of Brazil, and eventually Manaus, where we performed in the legendary Opera House built by rubber barons in the late 1800s.

I customarily rose at dawn intent on seeing some birds despite the fact that the shoreline was sometimes thirty miles away. One morning I noticed that our ship was barely moving and saw our Greek captain outside the pilothouse scanning the water. I asked him what he was doing. The Amazon was a tricky river, he said, full of shifting shoals and flotsam. He was looking for the Pink Dolphin, which always swam in the deep water. Within minutes we saw the magnificent Boto as it lifted like the arc of a rainbow out of the muddy depths, leading the way to safe passage. They are listed as a vulnerable species today because of pollution, increased boat traffic, and hunting. Bolivia has given them protected status as a national treasure and perhaps Brazil will do the same.

The animal Márcio Ayres chose to study as a young man was the rare and eerie-looking Bald, or White, Uakari. This primate, which lives only in the Mamirauá reserve area, has been called "man of the forest" by local tribes because of its resemblance to human beings. It has long shaggy pale beige fur and a bare red or pink face that gives it a distinctly human look, an odd-looking man with a bad sunburn. Sometimes as many as a hundred band together, but we saw only a few peeking at us from thick leaves high above, before they effortlessly leapt sixty feet to the next tree.

Márcio sought to protect these unique animals from hunters who were driving them to extinction. At that time his own body was ravaged by cancer, but he lived to see the Brazilian government establish the Mamirauá Sustainable Development Reserve for the animals and the people who live there. It is protected by the International Ramsar Convention of the IUCN (International Union for the Conservation of Nature), the group of scientists that determine the most vulnerable species annually through its Red List.

The Brazilian tribesmen who first populated the area more than a hundred years ago for rubber extracting have adapted like the animals to the flooded forest, lashing their homes to pontoons and making a living from sustainable fishing practices and handicrafts. There are about six thousand of them in the twenty-two-square-mile reserve; they are guardians of the forest and river, keeping alive the legacy of Márcio Ayres, who died in 2003.

Not all stories of rivers on earth have a happy ending. Most of the great rivers are stressed from pollution, overdevelopment of their shores, damming, and global warming. The Amazon, the greatest river on earth because of its sheer volume, is in no danger of drying up in the near future, although climate change will reduce the number of its feeder streams as drought prevails. The river itself will increase its volume as rising sea levels push more water inland hundreds of miles west from the Atlantic Ocean, flooding the city of Manaus and beyond. Forests will be wiped out in its wake, increasing carbon emissions by default. Trees are the number-one mitigation of the acceleration of carbon emissions into the atmosphere, and Brazil's Amazon rainforest has most of them. Despite Brazil's commitment to protection of Amazonia, holding it together is one of the greatest challenges of our time.

16

East Africa

I keep a primitively carved ivory chain necklace in my bureau drawer. It is something I inherited from my Nova Scotia grandmother. It looks old, perhaps from the 1700s, when my family settled there. I think it is Elephant ivory and was worn by one of my French ancestors before she traveled across the ocean to become a farming Acadian; but for all I know it could be made from whalebone or even Walrus, a piece of scrimshaw from the seafaring maritimers of my family. In any case it would not get through customs today . . . if they found it.

In 1989, Richard Leakey, who headed the Kenyan Wildlife Conservation and Management Department, had the idea of burning the country's confiscated Elephant ivory to bring the plight of poaching to the world stage. Twelve tons, artfully piled by a Hollywood pyro technician and lit by Kenya's president, Daniel Arap Moi, went up in flames, representing two thousand dead Elephants and worth $3 million on the black market. Iain Douglas-Hamilton, who has dedicated his life to saving Elephants, said it was "a tiny fraction of what was killed." President Moi said, "To stop the poacher, the trader must also be stopped, and to stop the trader, the final buyer must be convinced not to buy ivory." It reminded me of Nancy Reagan's "Just say no," her charge to users during the drug wars of the 1980s.

The poaching did not stop. The trading did not stop. The demand did not stop. The desire did not stop.

In 1995 another fraction was ignited in Nairobi National Park—the gray ashes of another few thousand tusks, the bloody carcasses long reduced to bones following the gorge of jackals and vultures. These same bones lay on the vast plains and were sometimes found by Elephant families, recognized as one of their own, and fondled in a ceremony of grief. How do they know this architecture of the body? Elephants may never forget, but their embrace of bones is uncanny.

This majestic mammal has suffered greatly at the hand of man: killed, enslaved, their ancient paths obliterated. I watched a lone bull elephant stare at the chain link fence butting up against the city of Nairobi along his ancestral migratory route, a skyscraper jutting into the blue a mere one hundred feet distant.

Poaching and loss of habitat have reduced the current African Elephant population by thirty thousand annually since 2010, about ninety-six Elephants every single day. A continent-wide aerial census estimates about 500,000 animals out of many millions are left, but no one really knows how many still roam the forests and plains.

It is not possible to comprehend the loss of Africa without knowing what *was* Africa, and in many places still is. The continent blessed with the greatest number of megafauna on earth is experiencing unprecedented threats to those animals today from many sources. It is a war on wildlife, and there are warriors in the battle fighting to keep what is left and restore what was. The glory still exists in places, and the resilience of animal life coupled with the will of people make the dream possible.

Our twin-engine plane broke through the fog and rain, scattering a herd of zebras, and landed on a patch of earth on the Serengeti. Nothing in my life experience prepared me for the great plains of East Africa—not all the reading I had done, the pictures I'd pored over, or the dozens of movies and TV shows I'd seen about wildlife. I stood on that plain that stretched for hundreds of miles—in any direction that I turned it just kept going, to the horizon and beyond—and saw thousands of animals everywhere. There were Lions and Zebras cheek by jowl, Wildebeests, Ostriches, Buffalos, Hyenas, and Gazelles. Grazing in the branches of nearby acacia were Giraffes, and Elephants walked in stately procession toward a mudhole. It was the vastness and abundance I could not comprehend.

It came to me that my entire understanding of the Serengeti until that point had been circumscribed by the rectangles of screens or books or photographs, that my vision had literally been narrowed about three hundred degrees. Here the gentle curvature of the earth extended to the horizon and I at its center circled slowly in place as the magnitude of the experience homed in on me.

It was here in this part of the Great Rift Valley that early man evolved and eventually walked out of Africa to people the earth. His life was inextricably linked to the animals that thrived here, many of them the same species seen today. He must have followed the northward migration of the herds across the plains and back again south to the Ngorongoro Crater in the June winter of the Southern Hemisphere. He made killing tools out of stone and flaked the stone for scraping hides. These tribal families lived like the mammals they hunted and scavenged, as another mammal wandering the land in search of food and shelter.

Our band of wanderers in 1996 traveled by jeep along the same routes and captured animals on celluloid when we found them. The Wildlife Conservation Society was one hundred years old; trustees and staff made "Footsteps Across Africa" in a centennial safari. There were so many of us that we broke into pods of fourteen with a guide, several drivers, and a WCS biologist as our leader. Our pod included Howard Phipps, chairman of the board, and President William Conway.

Bill Conway specialized in flamingos and penguins, traveling annually to Patagonia to tag penguins and encouraging many to join him. His ardent conservation ethic invigorated trustees like me. The society had a long history of advancing wildlife conservation, beginning with a small herd of American Bison brought to a meadow near the Bronx Zoo in 1905 on what today is Pelham Parkway, a main artery into New York City. This act probably saved the Buffalo from extinction. The most prevalent hoofed animal on the western plains, gunned down to a handful, successfully bred in the Bronx and was reintroduced to protected areas in western states in ensuing years.

As president of WCS, Bill Conway expanded on the mission of those early years by envisioning zoos primarily as conservation organizations and breeders of endangered species, not solely collectors. The name change in 1993 from the New York Zoological Society was consonant with the mission.

On the Serengeti trip in 1996, we tented on the plains, not in the little pop-up numbers I was used to back home on family hikes but in luxurious, canvas-roofed rooms with bona fide beds to fall into. An Elephant could have flipped the entrance flap and stood full height in our tent, it was that big.

An even larger tent was the gathering spot for meals, the long table draped in white cotton and laid with cutlery, stemware, and candles in the evening. We lived like sultans, engaging in glorious adventures during the day and in stimulating conversation at night, like-minded men and women eager to find ways to preserve the natural wonders we were privileged enough to see. The scientists accompanying us knew how to do it.

We set out at dawn, three or four to an open jeep, riding along worn dirt paths on the plains. The biologist Patricia Moehlman pointed out a Golden Jackal couple, mated for life, with two small pups playing in the early morning light, their prominent ears tipped with gold. One had a run-in with a Wildcat, which arched and hissed as any house cat might before slipping down a hole. Zebras, Wildebeests, and Thomson's and Grant's Gazelles grazed on grasses nearby in a scene as bucolic as an Iowa farm.

A Brit named Marian studying Spotted Hyenas said she hated *The Lion King*'s depiction of them as skulking and malicious. In fact, said Marian, as we watched a mud-splattered clan of them on their scrapes, they were fascinating animals dominated by a complex matriarchy. Male Hyenas are subservient not only to the matriarch but also to all the other females of the large clan. The female even has genitalia that resembles a penis but is actually the birth canal for her cubs. The most common and opportunistic carnivore in Africa, the Hyena devours 95 percent of any prey it scavenges or kills, crushing bones and digesting hair with ease. The Maasai people even leave their dead to be consumed by Hyenas, they are so efficient.

The Wildebeests were calving in late February, stopping briefly to give birth and waiting just a few minutes for the baby to rise to its wobbly feet before moving on—a precarious moment that predators knew was opportune. A little one struggled to rise, the silvery slime of its mother's womb still clinging to its matted hide. The immense herd moved on, maybe a thousand animals following the grass newly sprouted with the day's rain. A gap of fifty feet soon separated the mother and her calf, exposing his vulnerability, and within minutes three Hyenas began to circle the helpless baby. The mother moved next to him but did not nudge him or make any sound we could hear. The calf tucked its spindly hind legs underneath and attempted again to stand but fell back on its haunches for perhaps the fifth time. His mother looked toward the retreating herd, seeming to deliberate the odds. She went a few steps toward them before she turned suddenly and raced toward the encroaching Hyenas, striking out at them with her front hooves. This stopped them momentarily, and just as it looked like it might be the end for the newborn, he staggered to his tiny hooves and trotted to the herd as if he were an old hand at the game. Instinct made sure he was. He and his mother were soon safely enveloped in the mass of grazing dark bodies.

Large birds strutted the plains: Ostriches, bustards, storks, and the tall Secretary Birds, looking like prim clerks out of Dickens, stepped with halting gaits across the expanse, while several kinds of vultures swooped in on carrion. The big Griffon, or White-headed Vulture, was the cutter, with a sharp-edged beak made for severing flesh from bone, the others benefiting from his work.

An amorous Kori Bustard, the largest flying bird in the world, took a shine to Conway, flagrantly fanning his tail, pooching his white

Me, Bill Conway, and WCS biologist Sebastian Chuwa have just had the thrill of seeing a Lammergeier, a huge Bearded Vulture, soar through a mountain canyon, Tanzania, Africa, 1996.

neck feathers, and strutting toward Bill with a come-hither look. He emitted a low booming noise when especially excited and Bill responded with a sexy growl, drawing the five-foot bird so close we thought he would hop into the Rover.

Bill was a fine companion. We shared a passion for birds and counted 250 species on this African trip, from colorful Lovebirds draped on branches like a Mexican candlestick to the Saddle-billed Stork, its spectacular red and black bill glowing in the sun shaft of a forest pool.

There are places in the world where more birds live, but nowhere on earth is there such a concentration of large mammals. Hippopotomi lay submerged in large ponds, Oxpeckers gleaning insects from their hides before water lapped at their feathers. Lions lolled on their backs in amber grass, paws dangling insouciantly in the air, while cubs tussled around them. A troop of Olive Baboons, bucolic in the shade of a tree, made a pecking order of grooming females lined up behind a big male who was looking vacantly at his toes propped on a log. A baby plopped into his lap and he gently stroked it before it scampered off. Two Giraffes swung their long necks wide to butt each other in the breast, a fight choreographed like slow-motion Kabuki theater. Seven East African Oryx, their straight black horns three times the length of their heads, grazed beneath an acacia tree.

Two Cheetahs were being harassed by a spunky gang of five Plains Zebras. The cats flattened themselves in the grass as the Zebras raced toward them, then popped up again and tore after the Zebras, which galloped off, only to turn around and repeat the sequence three times before the Zebras were satisfied they'd made their point and trotted off. It was a deadly game made tolerable by the element of play.

The rock outcroppings called kopjes where the Cheetahs hung out reminded me of a spot in southwestern Oklahoma called Quartz Mountain where Ed and I taught acting to talented teenagers at a program of the Summer Arts Institute. Impervious granite hills rose from an ancient seabed. There was a severe drought one year and the lake disappeared, exposing a blanket of red-cracked mud from shore to shore. We were walking across the bare expanse when a lone smooth rock caught my eye. It fitted comfortably in my hand. One side was sharp as an adze, and the park's naturalist said it was a scraping tool from ten thousand years ago, probably dropped by its maker in a deadly chase, because the tool was not finished. The migration of Bison and Antelope then, before they were decimated 150 years ago, was as extensive as that of Wildebeests and gazelles in East Africa, the largest mammal migration on earth today.

On the Serengeti it was not hard to conjure our ancestors seventy-five thousand years ago also migrating north to settle Europe and the Fertile Crescent, taking with them the story of East Africa's bounteous plains—the progenitor, perhaps, of the story of Eden.

"Serengeti" is the Maasai word for "the place where the land runs on forever." These nomadic tribesmen began pasturing their cattle on the great savannah only two hundred years ago. Earlier tribal peoples had traversed routes from the coast to Lake Victoria and Lake Tanganyika for thousands of years, later joined by Arab traders seeking slaves and ivory.

During the 1800s colonial hunters came to shoot the big five: Lion, Leopard, Elephant, Cape Buffalo, and Rhinoceros. They did such a thorough job of it that by the 1920s parts of the Serengeti were established as game reserves, regulating limits. Finally, in 1951, all killing was outlawed with the realization of how special the area was.

Serengeti National Park was established, adjoining the Ngorongoro Conservation Area. It is a UNESCO World Heritage Site and comprises 5,700 square miles of northern Tanzania. Three hundred fifty thousand tourists visit it and thirteen other parks annually, contributing 25 percent of the country's economy. The animals are clearly worth more to the government alive than dead.

Ngorongoro Crater is almost perfectly round, twelve miles across, with lush green grasses, forest outcroppings, and small lakes. It may be one of the most beautiful places to see wildlife in the world. We stayed in a lodge perched on the rim of the two-million-year-old volcano with a breathtaking view.

Out scouting for Leopards on the ridge rocks, our guide, Bjorn, took three of us to see a giant dead baobab tree for the owls. It was hollow inside and had been occupied by poachers in the past. Bjorn and I crouched down to slip through the small two-by-two-foot entrance while Ed and our friend Barbara lay on the ground to take a picture of us inside. The circular room was about twelve feet in diameter and dimly lit from above by the hollowed trunk open to the sky. There were no owls.

Ed urged me to move back against the far wall for a better picture. I pressed my back into the clammy bark, and suddenly the weight of a heavy body landed on my right shoulder. A large dark snake quickly slithered down my chest onto the earthy floor, beating a path for the entrance hole. "Black Mamba!" I cried, that being the only African snake I knew. Ed leapt to his feet, Barbara screamed and ran away, and Bjorn watched as the scared reptile disappeared into the grass. It was indigo blue, not black, with a light blue iridescence, and about five feet long—not a deadly Black Mamba at all. Even the Black Mamba is not all black, only on the inside of its mouth. But it made a good story. At dinner Conway quipped that I could have been lost to a snake, and a few days later I overheard a stranger speak of the "twelve-footer in the hollow baobab that almost got her."

Achuar tribesman Simon Santi paints his face out of respect
for the rainforest, Ecuador, 2014.

A Poison Dart Frog we spied
on a tree fungus, Kapawi,
Ecuador, 2014

These Salps may
have existed for
500 million years,
but with a central
nervous system,
they are the same
phylum, Chordata,
as human beings.

The Comet Moth in Madagascar is one of the largest, most delicate moths in the world.

A young Mahout with his charges in India, 2004

Gabriel changing the radio collar on the Tree Kangaroo Trish,
Huon Peninsula, Papua New Guinea, 2010

A chameleon grabs lunch, Madagascar, 2015.

The Norwegians in the embrace of a giant kapok tree

The Giraffe Weevil, Madagascar, 2015

This wild two-month-old Black-backed Gull adopted us in the summer of 2013 and formed a relationship with my little dog, Drama.

The killer in my Nova Scotia woods. This Merlin decimated the songbird population during the summer of 2014.

Me in the Bahamas

The light is always changing over the ocean.
The view from our Nova Scotia home

Modern Elephants evolved around 500,000 years ago. They missed dinosaurs by about 64 million years, but as the largest land mammal on earth the Elephant awe factor is considerable. We saw them daily in Ngorongoro. One herd descended the crater rim at dawn in single file through alpine flowers and a stand of cactus, silhouetted against the rising sun, looking like a calendar shot, except there was no frame. A nervous male confronted us another time. He was twelve feet tall at the shoulder with big tusks, and wanted to pass our Land Rover on a wooded road. He flapped his ears and trumpeted and just when we thought he might charge, he dashed by into the brush.

One evening we saw more than a hundred Elephants in two herds, some with little babies—at least three newborns, one or two days old, who barely came to their mothers' knees and were tenderly watched over by aunties and siblings. Older ones were frolicking in the mudholes and chasing and squirting water all around. We watched for more than an hour before a nervous guardian edged toward us with ears flapping. Two others began running in our direction. It didn't look good, but they passed by us at a healthy clip and we noted the pursuer's giant semi-erect penis as he chased the pretty little female.

The Elephant is known for its intelligence, its affection, and for never forgetting. Lawrence Anthony, writer of the bestseller *The Elephant Whisperer,* was beloved by all for his rescue and rehabilitation of animals harmed by human atrocities and was known for saving the zoo animals in Baghdad in 1993. On March 2, 2012, he died suddenly at his home. Two days later a stately procession of two groups of wild Elephants, thirty-one in all, arrived there. They had traveled in single file from twelve miles away. They had not visited him for years and they stayed for two days and two nights without eating and then turned and slowly marched away. How can one not be moved by the inexplicable arrival of these remarkable creatures? A great-hearted man had died, and it seems they did not forget him.

Another uncanny story was told to me by a WCS biologist studying family structures of the animals. A young bull Elephant entered the lab tent and plucked a large femur bone off a shelf, took it outside, where he fondled it for several minutes, and then gently placed it on the ground and returned to the bush. The bone was that of his mother.

We know Elephants grieve, comfort, and console others in their herd. They seem to have extrasensory abilities as well. They are consistently rated one of the most intelligent mammals on earth, along with pigs, dogs, dolphins, chimpanzees, and orangutans. Throughout history we have exploited Elephants' brute force for our use, neglecting what we might have learned from them as supreme social beings.

They were used for hauling as early as 4500 BC in Mesopotamia until they became extinct there by 850. Alexander the Great had one hundred of them in 326 BC as he fought his way across India, although he could not compete with the legions of thousands that some Indian kings presented on the battlefield. Elephants were the first tanks in warfare, impervious to arrows and even musket balls. When the cannon was introduced in the 1400s, Elephants could be felled by a single cannonball, and their warrior days were essentially over.

The Elephant is still a beast of burden in parts of Asia, but as the wild population declines, so does the likelihood of taming one. As the land is squeezed by an increasing human population, feeding and housing a big Elephant becomes prohibitive. Ringling Circus no longer has Elephants in their shows; animal rights organizations finally convinced enough people of the cruelty involved in training an Elephant that the circus gave in. Even zoos are rethinking their Elephants. The Bronx Zoo is phasing them out. As the complex social lives of the animals became better understood, the zookeepers were reluctant to split up family groups and did not have the space to care for all of them adequately. Its last three Elephants will live out their long lives, and then one of the greatest zoos in the world will close its doors on the most iconic wild animal on earth.

I was of two minds about this decision as a trustee. While in no way endorsing splitting a family apart I argued that there were many domesticated orphan Elephants who could use a home and give the public the opportunity to know a live animal through the zoo experience. The little girl in an apartment in the Bronx may never get to Africa or Asia to view one in the wild as I have been privileged to do, but at least she can come to know them at the zoo. The first introduction most people have to wild animals is in a zoo, making a direct visceral connection to the possibilities of experience in the wild and protection of those creatures.

Without the zoo experience I worry that any media connection serves only to distance our children from the animals, which, after all, live forever on the screen even if they are gone in real life. Richard Louv in his book *Last Child in the Woods* calls it "nature-deficit disorder." Children today are joined at the hand to their tech devices and imprinted on visual imagery from birth. They prefer to be inside where the plugs are and engaging in virtual reality rather than in actual reality.

Elephant ivory has been coveted as long as human beings have been artists. When I was a girl wandering through Boston's Museum of Fine Arts I would gaze at a little statue of the Minoan Snake Goddess made of ivory and gold. She dated from 1500 BC and was exquisite in her tiered gown, holding two writhing cobras in her hands. Three thousand years later the trade in ivory increased dramatically as tusks were made into piano keys, billiard balls, boxes, religious objects, statues, and jewelry. Even my forebears couldn't resist the little ivory necklace I inherited.

With the added pressure of sport hunting the Elephant population began to crash. By 1978 the population had decreased from 26 million in 1800 to 600,000 and the Elephant was listed as "threatened" by the U.S. Endangered Species Act. This could have been a turning point—the species might have rebounded—but the Convention on International Trade in Endangered Species dithered about banning all trade in ivory, overruled by countries for which it was too lucrative a commodity. By the late 1980s a growing middle class with money to spare in China, Thailand, and Indonesia began to purchase huge amounts of ivory objects and the demand skyrocketed.

Another turning point came in 1989 when the first pyre of ivory was burned in Kenya, bringing world attention to the loss of Ele-phants, and CITES finally banned all trade in ivory. But just ten years later, in 1999, CITES allowed three countries in Africa to sell fifty-five tons of ivory to Japan, resulting in both legal and illegal ivory markets. Forged documents easily convince buyers that the ivory they are purchasing is legal and not contraband, and the level of corruption from high government officials on down, from the enforcers to the local spotters of big tuskers, is all pervasive.

There were many hand signals flying across the table as the tall, engaging men and I tried to communicate. There was a lot of laugh-ter, too, as we failed. Our hearts were light and there was much to celebrate. The Maasai were about to manage a new park in Kenya where they had grazed cattle for centuries. The former WCS biolo-gist David Western, Leakey's successor as president of Kenya's Wildlife Service, managed something of a miracle by convincing conservationists to put the new park in the hands of the Maasai.

Kenya was devastated by poaching in the 1970s and '80s, and the rising human population continued to press park edges with more farms, water diversion, and wildlife-human contact. The Maasai,

Mount Kilimanjaro and Elephants, Kenya, 1996

angry that they had been forced out of their ancestral pasturelands when the parks prohibited them, retaliated by killing Elephants, Lions, and other wildlife. But the Maasai tribesmen had managed the land well for centuries, pasturing their cattle in the habitats of Wildebeests and other ungulates without detrimental effects. David Western worked with them and convinced others that the new park should be run by the tribal people. WCS supported David's programs, and all of us were invited to a celebratory dinner in the center of the new park, with Mount Kilimanjaro's white peak glowing in the sunset beyond us.

The Maasai men had changed from their tribal reds, which drape gracefully like Roman tunics, into the monochrome navy jackets worn by Western businessmen. Toasts went on for hours, and we consumed the beef before us dutifully. The teeth of the Maasai were so perfect and straight that I entertained the thought of becoming a convert to their diet of beef, milk, and blood in an effort to transform my own. After hours of toasts and a presentation to our chairman of a spear and a shield, we danced together to drums vibrating on hard-packed earth until the moon was halfway through her orbit.

By 2100 half the children in the world will be African. That is UNICEF's projection. The population will grow from 1.1 billion in 2014 to 4.2 billion. This is an unprecedented rate of growth and will impact the entire globe. Animals will continue to be stressed. In 2009 a severe drought caused many ungulates to perish or move out of the park to areas where they could find grasses, leaving carnivores with little to feed on. Lions became cattle killers and the Maasai herdsmen poisoned them. Half the Lions died. In 2010 rampant poaching began again as the price of ivory in Asia escalated. Ten thousand Elephants were killed in Tanzania alone.

David Western and Cynthia Moss, another ecologist dedicated to Elephants, managed to create a safety zone for the animals to migrate freely across Tanzania's and Kenya's borders. It spans sixteen parks and protected areas, securing corridors through which Elephants can move beyond the parks. Elephants are not welcome in agricultural lands, where they tear up crops and trees and sometimes kill people. Western and Moss began programs to reduce human-animal conflict and educate people about the benefits of wildlife. Elephants open forests to grasslands, dig to create water holes, and disperse seeds in their dung. The tourist trade ranks second economically.

Independently, a young Maasai boy in Kenya, Richard Turere, charged with watching over his family's cattle as all Maasai boys between the ages of six and nine are, came up with a simple solution to deter Lions. They were getting into the bomas, or enclosures, at night and killing the livestock, and then were killed in return. Richard observed that Lions shied away from flashing lights. He took apart his mother's solar-powered radio and installed flashing lights on the posts of the boma and the Lions retreated. The concept was so successful that it was replicated elsewhere, and in gratitude to Richard, he received a scholarship to a fine school. He also was flown to the United States, where he gave a charming TED Talk in February 2013.

Leela Hazzah grew up in Egypt longing to hear the roar of a Lion before learning from her father that they had long been extinct in North Africa. Young Maasai men traditionally hunt and kill a Lion as an initiation rite to become "warriors." This was unsustainable in a Lion population that had plummeted from 200,000 to 30,000 in twenty-five years. Leela knew what she wanted to do from then on. Her Lion Guardian project is an unmitigated success with the Maasai. She convinced them that the more heroic task was being a lifelong guardian of a Lion rather than its killer. The Maasai warriors now use telemetry and cell phones as well as traditional scouting to guard their Lions. They are paid about $100 a month to be a guardian. The program extends throughout Maasai lands in Tanzania and Kenya and is resulting in a rebound of the cat population. Leela has heard the Lion roar.

Ivory is beautiful, it is valuable, and it is deadly. Elephants could be extinct in the wild within a decade if the carnage continues. The solution is to ban all ivory trade, to burn all existing ivory, and to put all the criminals behind bars. But different countries want different things and CITES has not found consensus on the issue yet.

Desire is the hardest thing to kill. The United States has been fighting illegal drugs for as long as I've been alive, but the demand continues. "Just say no" is better suited as a phrase on a refrigerator magnet than as policy. Curbing desire is a cultural phenomenon. It results from enlightenment coupled with shame, a realization that something is no longer cool to do or acquire. The desire may go underground and a black market may thrive, but cultural changes will have occurred in society, cutting demand considerably.

This happened with cigarettes when enough pictures of black and diseased lungs were imprinted on the minds of millions of Americans, and the surgeon general in 1964 reported that smoking was killing us. Forty-two percent of the public smoked then, compared with eighteen percent fifty years later.

It also happened with the fur trade in the U.S. No one thought much about the animals being killed for fur coats back in the 1950s. Anyone who could afford a fur coat was stylish and warm in the winter. The more exotic the animal, the luckier the wearer. First Lady Jackie Kennedy sported a Leopard coat in 1962, beginning a craze for spotted-cat coats. Hundreds of thousands of wild cats were killed—133,064 Ocelots in Central and South America alone in 1968. A cutter for the fur trade remarked that it took twenty-five of the small cats to make one coat, and that "they must be killing these animals off very, very fast . . . they're flown here with the blood still on their fur."

By 1968 the cats were becoming scarce in the wild and the price of a fur coat escalated. The anti-fur movement led by People for the Ethical Treatment of Animals began in earnest then and was fought vigorously by the fashion industry. Bill Conway wrote, "With what right and what conscience can a civilized woman adorn herself with the mummified reliquiae of diminishing wild creatures? How can she help but see an ugly death in a far-off land, the loss of one more portion of the international resource in beautiful wild creatures, each time she dons her 'fun fur'?"

The next year, 1969, the U.S. Congress enacted the Endangered Species Act. The plight of wild animals was part of the cultural conversation. A decade later most fur-bearing animals hunted for the trade were listed as threatened or endangered by the IUCN's Red List, and contraband because of CITES regulations.

Although this did not end the fur trade in farmed animals or the fashion industry's annual parade of top models in dead mink, chinchilla, or fox, it did reduce the number of animals caught in the wild and succeeded in shaming women in prominent cities of the world. With the invention of new fabrics that are fashionable and warm, there is no need for anyone today to deck themselves out in skins.

Even in South Africa, Isaiah Shembe's Nazareth Baptist Church, which follows old Zulu practices relying on Leopard skins for ritual ceremonies, is making the switch to faux fur with the help of Panthera. It looks lush and is easier to come by. The Leopard population is rebounding.

The connection between ivory and dead Elephants is being made in the places where demand is highest. Yao Ming, one of China's wealthiest and most beloved celebrities, played for the Shanghai Sharks before being recruited by the NBA's Houston Rockets. Now retired and living back home in China, Ming has become a voice for endangered wildlife. He led a campaign to stop shark finning, and the consumption of shark fin soup in China dropped by half. He has taken on the decimation of White Rhinos and Elephants. He made commercials telling people not to buy ivory and appeared in films such as Animal Planet's *Saving Africa's Giants with Yao Ming*. In one video an adorable baby Elephant, orphaned through poaching, follows the seven-foot-six-inch Ming through the African bush.

The Chinese film star Li Bingbing is one of China's top actresses. She visited Kenya in the spring of 2013 at the invitation of Iain Douglas-Hamilton and the Save the Elephants Foundation. Iain said that when she witnessed a slaughtered Elephant, its face torn apart for the tusks, she wept uncontrollably. She confesses that she had not made the connection between Elephants in Africa being killed for the ivory and the jewelry she purchased in China. Many sellers tell people that the ivory is from the dropped tusks of live Elephants, or ones that died naturally. Bingbing has become one of Asia's leading advocates for Elephants, educating youth and changing the minds of ivory buyers everywhere.

It is because of young conservationists like Yao Ming and Li Bingbing that Douglas-Hamilton says he has hope for the future. He has spent fifty years studying Elephants and has seen the horror during aerial surveys of the decimation of entire herds. He has witnessed the near extinction of Forest Elephants poached by warlords; Joseph Kony has used the sale of ivory to fund his Lord's Resistance Army (LRA). When a deranged man like Kony has no conscience capturing 66,000 children to serve as sex slaves and soldiers in his "army," the killing of thousands of Elephants means nothing. He and other terrorists are equipped with the latest in technical devices and weaponry to track down and kill Elephants and Rhinos and elude detection. It is an all-out war but a war that can be won with the will to do so. Secretary of State Hillary Clinton first saw a security threat to the United States and Africa in Kony's LRA becoming rich on the ivory trade. She and Chelsea Clinton, with funds from the Clinton Foundation, have committed to ending all wildlife trade.

While Yao Ming, Li Bingbing, and conservation organizations like WildAid work on the buyers of ivory, well-armed enforcers in anti-poaching squads work to eliminate the sellers. In 2012 two jewelers were busted with $2 million worth of ivory, uncovering a large underground market in New York City, second only to Hong Kong's in size. There are only eleven Fish and Wildlife inspectors at JFK airport and at Port Elizabeth in New Jersey and they have less than a day to inspect each huge container before the packages are shipped out. This shrinks the likelihood of nabbing most illegal trade in wildlife and body parts. Still there is room for hope. President Obama banned all trade in ivory in the USA in 2016, spurred on by very active organizations like WCS. More laws are sure to follow.

Sometimes I wonder if we have not conjured Elephants. They seem mythical and sprung from our imagination with their strange long trunk, their huge feet and ears, their impermeable hide, their small eyes, and their flexible tail. We might expect them to be un-gainly, but they are delicate and precise of movement, gentle except when pushed to anger, forgiving instead of vengeful. Clearly their minds are superior—housed in an odd huge body.

I like to believe we are transitioning our thinking. Elephants will not be beasts of burden much longer, not when villages have machines to do the work. The poaching will end when there is enough enforcement to stop the killers, as Nepal has proved with their army at the ready, and if funds are there to ensure it. There are many ifs. If Iain Douglas-Hamilton has hope, then so do I. He has seen it all and he says the Elephants are still forgiving after so many decades of slaughter. In addition to their remarkable intellect, they have very big hearts.

17

Madagascar

Madagascar existed like a giant seed held within the pod of Gondwana, supercontinent of the Cretaceous period. It was pressed on the northwest by Africa, on the south by Antarctica, and on the east by India. About ninety million years ago the landmasses separated, and Madagascar's flowering in isolation began. It is one of the most fascinating places on earth.

The fourth-largest island in the world, Madagascar also ranks as one of its poorest countries. Most Malagasy people earn less than $2 a day. The main road traversing the center of the island, a thousand miles from north to south, is riddled with potholes the size of refrigerators, making travel slow. It allows one to gaze at the beauty of the landscape and the industriousness of the people, who virtually live outdoors selling their vegetables or charcoal by the side of the road. The juxtaposition of vibrant green rice paddies terraced up the mountainsides with red-clay earth and the tidy brick homes made of it is stunning. This beauty comes at a price. The island used to be covered with forest.

Two thousand years ago the intrepid and most skilled sailors the world has ever known, the Polynesians, colonized Madagascar, as they had thousands of islands in Indonesia far to the south and east across the Pacific. When they arrived they found huge animals: the Elephant Bird, related to the Ostrich but ten feet tall and producing an egg equal to 180 chicken eggs, a lemur as big as a gorilla, giant tortoises, and crocodiles. As with large animals anywhere they did not survive this human invasion and disappeared within a few hundred years.

Madagascar's long isolation as a landmass produced more unique creatures and plants than anywhere else on earth. These endemic species are almost all threatened with extinction today as the human population continues to expand, obliterating the forests on which so many of the species depend. Only 7 percent of the original cover is left.

The island is not electrified for the most part, so people rely on charcoal for heat and cooking. The Malagasy depend on rice, eating it three times a day. Paddies, although basic and beautiful, are extensive as they replace forest after forest, and more lemurs, chameleons, and boa constrictors lose their home.

There are more than a hundred species of lemur. More are being discovered every year even as some are blinking out. They occupy specific niches: some high in the canopy eating flowers or fruit, some low and nocturnal eating leaves or insects, some leaping across the grasses like so many acrobats let loose on a holiday. They are comical like the bouncy Ring-tailed Lemurs, mournful like the large Indri, whose high-pitched whoop rings through the forest, or scary like the Aye-aye, which seem dressed for a Halloween fright night.

My twin grandsons Mac and Finn were nine when they decided where they wanted to go on their twelve-year-old trip with Nana. Finn said he needed to see where Nelson Mandela had been imprisoned for so long on Robben Island, South Africa. Mac wanted to see Lemurs and other endemic species in Madagascar, fearing they might not exist when he was older. In 2015, we set off together on a great adventure while they were still eleven, just short of their twelfth birthday.

Robben Island is within view of downtown Cape Town—a long spit of low scrub and sand surrounded by shark-infested waters that ensured prisoners could not escape. Apartheid, the policy that segregated blacks and kept whites in power, is as dark a legacy of South

Africa's past as slavery is of America's. What is celebrated, however, is the grace and fortitude of one man, Nelson Mandela. Being confined to a small cell, sleeping on a thin mat, laboring in a quarry for eighteen years on the island, and nine years elsewhere, forged a giant of a man who rose above despair, embraced nonviolence, and changed his country forever. Should we forget, in the tumult of violence that embraces eras like our own, that human beings have greatness within them, we need think only of Mahatma Gandhi, Martin Luther King, or Nelson Mandela to know what is possible for our species.

Tom Moses was a political prisoner for many years on Robben Island. Now he leads tours through the very rooms where he slept and labored. His story of survival, forgiving his jailers, and returning to the island to tell his tale was deeply moving. We were quiet on the boat ride back to town.

The next day we drove farther down the coast to visit a colony of endangered South African Penguins grunting to each other in courtship on the beach and tearing grasses and twigs from the sands to line a depression for nesting. In a bay farther south, Southern Right Whales and their calves fed on plankton before their summer migration. And even farther south, in Gaansbai, the boys donned wet suits on a shark-diving boat and were lowered into the water in a cage just seconds before a Great White Shark careened through the murky water inches from their faces. "Awesome!" they shouted, giving the thumb-up sign. The ten-foot shark was not after the boys; she was after the chum thrown into the water to lure her near. Nana stood on deck taking pictures. The marine biologist said that little is known of their breeding patterns and where they rear their young. Because it is a species in deep decline it is vital to learn as much as possible.

We could have spent another week in Cape Town, but we had a date to see the great mammals of the world in the Bataleur Nature Reserve bordering Kruger National Park. We were alone in the small, neatly tented camp along with our guide, Wayne; the tracker, Doc; and my friend Gary Allport and his son Jude, who joined us

An Emperor Moth with eyes on its back

Giraffes and Zebras keep watch together.

A Leopard sleeps with a full belly.

A giant millipede

A baby Hyena looks at us curiously.

from Mozambique. Gary is senior adviser to the CEO of BirdLife International, which represents 120 countries. I met him as a member of the BLI's advisory group. He is an ace birder, and Jude at nine was well on his way to becoming the same.

Our days were spent in an open Land Rover roaming the vast scrub savannah watching Elephants browse leaves and tear saplings down to get at them. Giraffes at a higher level extended their great tongues, curling them around green acacia branches. Zebras and Impalas stood like security guards alert to encroachment by Lions and Leopards. A Warthog crashed our picnic lunch one day and Wayne, easy and respectful, quietly shooed him away before he was able to snatch a sandwich from the table.

We were fortunate to see eight White Rhinos, so heavily poached throughout southern Africa that they are threatened with total annihilation, as are Northern White Rhinos, now down to only three, all in captivity. One White Rhino horn is now worth as much as $80,000 on the black market.

The boys, for whom all moving things are worthy, scrutinized the smallest lives, those of dung beetles, and leaf bugs. The ten-inch millipedes crawled up their arms with the titillating brush of a hundred little feet, while a Black Mamba rose to half its height by the side of the road to peer at us. Doc taught us the tracks of animals covering the dusty paths and dry stream beds. We never saw a Lion, despite following tracks of a female and her cub. She was hiding him away for safety. But a beautiful Leopard lay sleeping against a large inactive termite mound in the evening light, her belly swollen after a meal. The peace of the reserve and the calm of the animals used to nonaggressive humans made it a kind of Eden.

We had a passion for birds and were blessed with numerous sightings of Hornbills and a couple of Ostriches. The best one, though, was the male Korhaan, or Red-crested Bustard, engaged in a bizarre courtship display that entailed shooting straight up in the air about twenty feet, turning over, spreading his black feathers like an upside-down umbrella, and dropping back down to earth. This outlandish behavior may have failed to impress the females, but we thought it was super.

The boys were in heaven. Their love of exploration and discovery was fulfilled at every turn, and after night rides capturing eye shine in our flashlights of owls, rodents, or a Wildcat, we would tumble into bed and be ready at 4:30 the next morning for another excursion.

We flew to Madagascar in the second week of our journey, encountering a totally different environment, hardly Africa at all. Over ninety million years the island evolved into a benign, almost predator-free atmosphere. There was only one cobra to fear, few poisonous insects or frogs, no poison vines, and nothing to threaten the lives of human beings except extreme weather. One could walk through the forest with ease. The lemurs feared only the Fossa, a catlike creature that preys on them, a few raptors, and some human beings who hunted them for food, pets, or pelts.

Patricia Wright is a lemur expert. She discovered the Golden Bamboo Lemur as a young woman doing research in the rainforests of Ranomafana in the south-central part of the country. We became friends when she won the Indianapolis Prize for her scientific achievements. The prize from the Indianapolis Zoo, given by the Eli Lilly and Company Foundation, comes with an unrestricted cash award of $250,000. This biennial event is like the Academy Awards for field biologists. It is a gala evening celebrating their accomplishments. I became involved as mistress of ceremonies the first year and was honored myself in 2012 with the first Jane Alexander Global Wildlife Ambassador Award for my promotion of conservation and field biologists. The JAGWAA I call it, noting to myself that I am never far from the love of great cats.

Pat met us at the airport and we began the nine-hour drive to the research institute Centre ValBio, which she founded in Ranomafana Park in 2008. It is a stunning modern building housing dozens of visiting scientists who come to work on all kinds of species, from lemurs to leeches. I was given a private room reserved for professors, while the boys bunked in a room on the floor where the researchers and grad students lived.

The "Bat Team," a group of scientists hailing from Spain, Finland, and Britain, along with local Malagasy, had just discovered a new bat species and gave a presentation on the importance of bats as insectivores, particularly for certain agricultural crops. The "Frog Team" also worked at night, suiting up in climbing gear to capture the tiny amphibians from the watery center of large ferns or high in the tree canopy on leaves or bark. These guys and gals, most of them in their twenties or thirties, were "very cool," in the words of the boys. They spent their days logging information on their computers and their nights roaming the forest with headlamps. I could see Mac assessing the possibilities of science as a career. Finn might gravitate to something more art related; he carefully composed every photograph. "All knowledge will serve you," my father used to say, no matter what the future holds.

My birding friends tell me you have to get a child before he or she is eleven if you wish to imprint birdsong and behavior deep within rather than having it pass through the intellect first. It is like learning a language. A first language is never lost. It can be revived many years later even when unspoken for decades.

I sometimes struggle with identifying songs and calls because I came to birding so late. I will never be a master birder, but I know enough to know how much I don't know. Mac was smitten when he was a baby. He would watch birds at our feeder before he could crawl, and one of his first words was "dub," for the Mourning Doves he saw.

I took him on his first Hawk Watch when he was two and a half. He gazed up at the sky, calling out "cwow," "hawk," "bwuejay," or "seagull" unerringly, to the surprise of adult watchers. His powers of observation were so acute that he had absorbed the look—the "jizz," as birders call it—of a bird, without needing to break down the details of color, feathers, or size to ID. Most of us can do this if we see loved ones walking at a distance—we can tell who it is simply by the way they walk, the way the parts of their body move together. The top birders identify birds this way, by sight and sound, and they began at an early age.

I watched as my grandsons absorbed all the new sounds and sights with ease and had the ability to recall them days later. I wondered how long it would stay with them, particularly when puberty hit and all focus shifted to raging hormones.

Our initial walk in the lush Ranomafana rainforest resulted in our first lemur sighting. High in the canopy was Patricia's own discovery, the Golden Bamboo Lemur, crunching on a bamboo stem and allowing small pieces to rain down on us. She was beautiful, her gold-hued coat glowing in a late afternoon shaft of sunlight. In the days that followed we watched a Red-bellied Lemur youngster torment his parents and siblings with bold leaps onto their backs and heart-stopping jumps from branch to branch. The Greater Bamboo

Lemur occupied a different niche nearby and munched on a different part of bamboo, making it possible for the different species to thrive together. Darwin would have loved this, I thought.

The tiny Mouse Lemur came out at night and was lured into view by bits of banana and mango placed on a branch. Its big eyes were caught in our flashlights for only a few seconds before it scampered away with the fruit.

Greeted one morning with a heavy downpour, Pat suggested we go south, where it "never rained" and Ring-tailed Lemurs held court. Huge granite boulders and bald mountains left over from glaciations dotted the landscape in the southernmost tip of the country. Anja was home to several troops of Ring-tails. We had barely alighted from the car when a family leapt across a field heading for a ripe mango tree. They were hungry, greedily pushing the soft yellow fruit into their mouths, taking little notice of our cameras. The boys inched closer and closer until they were just two feet from an inquisitive youngster who peered at them with equal curiosity: primate to primate.

The scene was bucolic. A herd of Zebu cattle knee-deep in a lake

A Ring-tailed Lemur, Madagascar, 2015

The lemur expert and anthropologist Patricia Wright

pulled on submerged grasses and the granite rocks rose majestically into the blue while lemurs draped the mango trees like so many holiday decorations.

We saw several of Madagascar's three hundred or more chameleons, fifty-nine of which live nowhere else. No chameleon, however, was as wondrous as the giant one in Ranomafana that measured two and a half feet from tip to tail, its tongue unwinding like a blob of bubble gum for twelve inches to flick in a grasshopper.

Pat's love of lemurs was contagious. She loved everything about Madagascar, but the expanding human population was destroying more and more forest in its effort to thrive. The people slashed and burned trees to create new rice paddies or taro crops, or to mine the many minerals and gems Madagascar is blessed with. Each forest fragment is home to some unique creatures, and when that home disappears, they disappear along with it.

As an anthropologist, Pat knows it takes a village to safeguard anything. While we were at Centre ValBio, Pat met with elders to discuss the illegal gold mining that was taking place in the park, cutting into the forest, polluting the waterways, and disturbing the wildlife. The perpetrators, whom they called "bandits," were outsiders armed with guns, a frightening situation for the unarmed Malagasy.

With the blessing of the people she helped protect Ranomafana Park by establishing rules to benefit animals but also to give employment to the Malagasy as guides, rangers, and foresters. She boosted the ancillary benefits of tourism through encouragement of local crafts like weaving and embroidery, eco-lodges for tourists, and better health, education, and welfare for the local community. When she is not teaching at New York's Stony Brook University, she is raising money from public and private sources for Madagascar and its wildlife.

Promoting the wonders of Madagascar is easy for anyone who is fortunate enough to visit even a small part of the land as we did. The problems of keeping it wonderful are ongoing. It is a country that needs an infusion of bold ideas in infrastructure: green energy systems, transportation, agricultural practices, and tourist facilities. Madagascar could be a model for the world, particularly in thinking how to address climate change. (Richard Branson and Bill Gates, are you reading this?) The Malagasy are bright, congenial, and hardworking. Without thoughtful development the country will topple toward bankruptcy—economically, yes, but also bankruptcy of the people's future and that of the unique animals and plants that inhabit only this fragile part of our world. We are all the poorer if they disappear—the magic that is Madagascar, the dream that is Africa.

PART 3

———

THE BODY OF THE EARTH

When Dad and fellow Harvard alumni physicians shipped out for Britain just a month after Pearl Harbor was attacked in December 1941, my mother, my baby brother, and I moved with another mother and her three small children to a house where we waited out the war years together. The house was charming, with a vine-covered trellis draping the side porch and a rabbit warren of rooms all the way to the attic. It was the sunken garden, however, and the panoply of diverse, stately trees—oaks, sugar maples, and a glorious magnolia shading the driveway circle—that struck me so vividly at age three. It is my first memory of beauty.

These were no ordinary grounds. Ninety-nine Warren Street in Brookline, Massachusetts, was the home of the great landscape architect Frederick Law Olmsted, who designed Central Park in New York City and countless other parks. He is called the father of our country's municipal parks and of the art of landscape design itself. He was also one of America's first conservationists who entreated the public to follow the lay of the land and native cultivation when gardening.

The great man was forty years in his grave when we moved into 99 Warren, but the firm carried on under the tutelage of his sons. A dozen draftsmen were bent over high desks in a long light-filled building attached to the second floor of the house, giving us children easy access. Our mothers told us not to interrupt, but we did it anyway, and the men welcomed the respite of our moppet company, giving us tomatoes, beans, or carrots from a vegetable garden originally planted by Olmsted himself.

It was wartime and we kids chipped in by rolling tin into balls for the effort and pressing yellow and white packets together to make margarine. The fresh tomatoes and beans were a luxury our mothers gratefully accepted.

Germany surrendered in May 1945. Our mothers were tense, praying their men would soon be home. The 5th General Hospital had been sent to Normandy to care for the thousands wounded at Omaha Beach and along the coast of France in 1944. It would be seven more months before Dad performed his last surgery on these brave soldiers and sailed back across the Atlantic.

On July 9, 1945, there was a total eclipse of the sun. We five kids had paper bags put on our heads with slits for our eyes. We stood in the courtyard in the noonday sun like baby ghosts gazing at the sky along with the architects and our Scottie dog, who was racing around in frantic circles as the dark enveloped us. The heat of the day gave way to a subtle breeze as the sun turned black and a shiny ring of gold encircled it. I thought the sun blew the wind to us at that moment as it extinguished its own light—my first memory of mystery. It was enchanting living at the Olmsted estate.

My first encounter with water was also memorable. Mom took us to Walden Pond on a steamy summer day. With water wings affixed to our arms we splashed in the cool sanctuary and she taught us how to dog paddle. The pond in 1943 still had reeds and lots of frogs as Thoreau described it ninety years before from his cabin on the opposite shore. Today it is a busy recreational beach with attendant algae problems, but for us it was clear and clean with little fish swimming between our legs.

So my provenance of conservation was visited on me from an early age by masters. I was lucky. Yet beauty is found in nooks and crannies everywhere: in a shaft of light across pavement in the big city, in the eyes of a child, in the spill of water over rocks, or in the cut of mountains on the horizon. It is there for us, even in the ugliness we've inflicted.

The theologian Sallie McFague, in her eloquent book *The Body of God,* writes that every finite particle of our planet is the body of God and that until we love our own bodies and all the bodies we share the planet with we cannot begin to love and care for the body of the earth.

The body of the earth is *scarred.* The body of the earth is *sacred.*

Rearrange a few things and the relationship to earth changes. If our world seems beyond repair, leading to the irrevocable loss of many species, possibly our own, rearranging priorities can put us on a path to a holistic, even holy, path of healing our planet and our own despair.

There are thousands of stories of people around the globe taking matters into their own hands. There are organizations amassing millions of people online to put pressure on legislators to halt drilling in the Arctic, the spraying of pesticides, or the killing of wolves, for example. I have listed many of the most important organizations and their missions in the appendix at the end of this book.

Because national governments lag behind the environmental needs of local people, communities from Fiji to Laos to New York's Hudson River are adopting best practices for fishing, clean air, and water. Women in Africa are planting thousands of trees, while children in South America champion the lives of endangered animals in the places where they live. Place is a space with memory. Everyone is imprinted with memories of places they lived, places we long to return to and keep as they were. Nothing stays the same, but it is possible to halt the wanton desecration of land, water, and air. It is not rocket science. It is commitment.

This blue marbled beauty the Earth is like no other, as photos of Mars, and now of Pluto, more than three *billion* miles away, attest. If we can send photos back from Pluto, we can do anything.

Human beings have indomitable spirit, and so do animals. The will to thrive is inborn. We adapt, we evolve, and we survive best in harmony. These last chapters are about extreme conditions, best practices, deadly ones, peaceable kingdoms, and the miracles of life on earth.

18

Desert

My fingers wound around three reins in each hand as six horses thundered across the Arizona desert, me sitting high above them on a rickety old stage coach. The sky was blue as blue can be, the space was wide open, and the thrill of controlling the power of six horses was incomparable. I was playing Calamity Jane in a CBS TV movie of that name, and although an accomplished wrangler hidden inside the seat box was actually holding the lead rein it did not diminish my excitement. The camera rolled, the Indians began galloping after me, and I felt free as the wind. I only hoped my hat would not fly off. "Cut!" yelled the director through an old-fashioned bullhorn and we turned the six-horse stage around and did it again. After the second or third take there was a close shot of me standing on my head on the running stage screaming like a crazy banshee, which frightened the Indians away. I really did stand on my head, but the stage coach was not moving. The long shot was done by a stunt double. Still, my commitment to shooting, trick riding, driving the six-horse stage, and crashing through windows with a tuck and roll earned me the trust of the stunt people, who were my idols. They gave me a big bronze belt buckle with "Stuntmen's Association" engraved on it at the end of the shoot, and an honorary membership. It was like winning an Oscar.

We were four weeks in the desert south of Tucson. It was hot.

There was a spate of days where the temperature stood at 114. Our vigilant wardrobe woman was busy plying us all with wet cloths around our necks and salt pills to keep us hydrated. She neglected to minister to herself and collapsed one afternoon of heatstroke. The hospital did its best to save her but, shockingly, she died within days.

The sun is our life and our death. It is the fire that makes all life possible, threatening us with annihilation at the same time. Icarus flew too high. The sun melted his wings of wax and he plummeted to earth.

In deserts I always feel some fear, as if waves of heat or fire were about to roll suddenly over the next dune or scrubby hill. I have been there when dust storms and sandstorms obliterated all sight, when my scarf covered all but the slit in my eyes through the sunglasses, and still we trudged on. I have been low on water and felt panic. Where is the water? How do animals survive?

The Sonoran Desert embraces Tucson and extends west to the Salton Sea and Palm Springs in California, and south all the way down the Baja Peninsula into Sonora, Mexico. It is the largest of our country's deserts, and the hottest. And yet the Sonoran has abundant life, plants, and animals that have invented ways to store water or become torpid until water comes to them. They are little miracles.

My favorite is the Spadefoot Toad, which has a bloated little three-inch body of greeny-yellowish mottling, big inquisitive eyes, and an appendage on its hind legs for digging—hence its name. It digs a burrow in sand or hardpan soil and then waits until it rains. The vibrations from the drops or thunderclaps bring it to the surface at night, where it immediately begins to feed on insects and seeks to mate. Males bleat like little lost lambs in the pools made by the rains, and females find them. The females will then lay as many as three thousand eggs, which hatch within twenty-four hours, even if the desert pool is eighty-six degrees. The tadpoles become Spadefoots in nine to fourteen days, the shortest time period of almost all amphibians. But theirs is a race against time. The pools are evaporating, and in order to maximize their progeny they hurry things along. They can eat enough in one meal to hold them for an entire year underground. Still, of the thousands of eggs laid, few survive to adulthood; they become meals for small mammals and birds, and the victims of dried-up pools.

This is an amazing amphibian and, like most amphibians, is best seen on a spring or summer night or before dawn. Amphibians are declining the world over due to factors not fully understood including fungal disease, pollution, habitat destruction, climate change, and poaching for the pet trade. The Spadefoot is holding on and is listed currently of "Least Concern" on the IUCN's Red List.

The Sonoran Desert is home to a hundred reptiles, including the most venomous and famous, the Gila Monster. This is another animal whose adaptation to desert life is remarkable. Like the Spadefoot, its chunky body efficiently stores food. The Gila's tail is a kind of storage locker expanding with digested meals of baby rabbits and other little rodents that come with the summer monsoon season. Its bladder holds an enormous amount of water so that the lizard can retire underground for a week or more fat and happy. The poison in its saliva is thought to have a hormone regulating this activity, which may prove useful to humans in controlling type 2 diabetes.

As I watched a chunky pink and black Gila Monster hunker down beneath the sand under the shade of desert scrub, two Collared Lizards a foot high raced by in tandem on their hind legs like Victorian gentlemen late for a party. I saw nothing in pursuit. Certainly not the Desert Tortoise, which lumbers across the parched land munching on flowers and leaves and tolerating temperatures of 140 degrees. Like the lizards, it stores moisture from these plants in its bladder. This is a species the U.S. lists as endangered, declining because of off-road vehicles and poachers.

Reptiles do not usually rank high when people think of the illicit trade in wildlife, but there is a vast network of collectors who will pay thousands of dollars for rare species like the Desert Tortoise. The Gila Monster was the first to be protected, in 1952. Even though it breeds well in captivity, it is still poached.

Bryan Christy in his book *The Lizard King* documents the mafia-like global trade in illegal reptiles, with a cast of characters out of a thriller. The majority of U.S. trade is legal and done through retail pet stores or the Internet. There are also three hundred reptile shows put on annually in cities across the United States. I love spending an afternoon wandering the booths and talking to the breeders even as I cringe at the tiny jars and boxes the herps—reptiles and amphibians— are confined in. While contraband is forbidden, I've witnessed under-the-table transactions of exotic frogs from South America and baby boas from Asia. Most of the sellers are legitimate, passionate about their herps, and adept at captive breeding, which has sky-rocketed in the past twenty-five years. Stars like Bob Clark breed albino snakes with intricate patterns rivaling art deco.

When I was twenty-two I kept a Green Iguana in my New York apartment. Iggy, as I inventively named him, was a fine pet. He didn't need to be walked, didn't bark or bite, and was a cheery celery green with some stunning turquoise scales; his eyes tracked me across the room until he got his dinner of fruit and lettuce. I wouldn't keep one today, but I understand the popularity of herps. In any case, being captive bred, they cannot be returned to the wild.

Reptiles are not the only denizens of the Sonoran to be smuggled. The saguaro cactus—home to the little Elf Owl, which holes up between the spines, and to the Cactus Wren, which weaves a nest on them—stands as a majestic column as high as sixty feet with as many as twenty-five arms. They can live for two hundred years, growing achingly slowly. A ten-year-old may be only an inch and a half high, and the first arm may not appear on a saguaro for seventy-five years. Even when their weight tops two thousand pounds, their roots reach only a few inches into the soil, braced by horizontal extensions as wide as the plant is tall. The Sonoran Desert is their only home. The cactus is protected, but Saguaro National Park has major highways going through it, a fast getaway for poachers, who can collect thou-sands of dollars on the black market. Many of the most desirable wild cacti have been implanted with radio frequency chips to tip off the thieves' locations.

My friends Karen and Ken have a small adobe home within the park boundaries, living as if surrounded by elegant Roman statues, or caryatids of the Acropolis. The oranges and pinks of the Arizona sunset wash over the cacti columns in an expressionistic orgy when the desert begins to cool.

At 4 a.m. I wandered among them by starlight, the faintest line of light on the horizon, when a thousand quail suddenly babbled through the stillness with the force of a thousand alarm clocks announcing the new day.

There are few deserts on earth with the biodiversity of the Sonoran. The topography rises in hills and slices into small canyons and arroyos that rush full with monsoon rains; sometimes it hardly seems a desert at all. With more than a hundred reptiles and amphibians, sixty mammals, two thousand plants, and four hundred birds, it is one of the biodiversity hot spots of North America—an incubator of desert evolution.

At the western end of the Sonoran is the Salton Sea, the largest body of water in California and a case history of the state's problems. I went there for the birds. I can't remember what my ultimate destination was; it could have been Arizona or New Mexico, as I've made movies in both. I shot the movie *A Gunfight,* with Johnny Cash and Kirk Douglas, in a New Mexican desert where the horse wranglers wrangled a dozen ferocious Diamondback Rattlesnakes out of old wooden buildings and into large cages. Their heads were as large as fists and they kept banging them aggressively against the wire top whenever anyone came near. But it was not the snakes I feared—it was that I couldn't see water anywhere. On my days off I drove to the Rio Grande and sat with my feet dangling in the river.

Wherever I was headed that day from Los Angeles, I remember spontaneously taking a detour south to the Salton Sea because of the White Pelicans, fully 30 percent of the U.S. population, and 80 percent of the population of wintering Eared Grebes. Driving down the I-10 I passed Ferruginous Hawks low in agricultural fields seeking rodents, and the agile Prairie Falcon high on utility poles eyeing small birds. Wintering Sandhill Cranes paraded through harvested rows picking up old seed, while flocks of Blackbirds careened overhead.

The Salton Sea is on the Pacific flyway, the migratory highway for birds heading south for the winter or north for breeding, as far north as the tundra of Alaska and Canada. It is the last big body of water on the southern trip before the Sea of Cortez, sort of like a service center on the interstate; you miss it and you may go hungry for the rest of the journey, running out of gas before you reach your destination. Millions of birds make a stopover, 450 different kinds, some off course and extremely rare.

There was not the searing heat I expected in one of the lowest points in North America, just a few feet shy of Death Valley's record 250 feet below sea level, but I was not there in midsummer, when the temperature of the water alone can hit ninety degrees. Hundreds of White Pelicans sat on old pilings and rocks or occasionally swam in lines in the water trolling their massive beaks like seining nets to collect fish, perhaps the introduced species Tilapia, capable of withstanding hot water.

The Salton Basin is an ancient inland sea that has expanded and contracted throughout history. The modern-day Salton Sea was born of an accident in 1905 when the Colorado River overran gates to the Alamo Canal as a cascade of heavy rains and melting snows dug into sands and spread for 370 square miles. On my visit in the 1970s the marina was still active and boats were buzzing about. By the late '90s the boats sat high and dry, hundreds of feet from water, and the marina was dead. The coup de grace will come in 2018 when a deal among the U.S. Department of the Interior, the State of California, and numerous water agencies takes effect. Water to the Imperial Valley and Salton Sea will be reduced for seventy-five years, until 2093, to allow growing cities like San Diego to receive what they need.

Meanwhile the Salton Sea shrinks, salinity increases, and algae blooms suffocate the oxygen, causing untenable toxicity. Dust from the dry seabed area is so toxic that people's health is compromised as far away as Los Angeles when the wind kicks up. There are die-offs of thousands of fish and the concomitant deaths of thousands of birds—twenty thousand Eared Grebes in 1994 alone. With no food and nowhere to go, they have come to the end.

Audubon chief scientist Gary Langham has mapped projections of habitat loss until 2080 due to climate change throughout North America. We may lose as many as 314 species because of extreme conditions such as those in the Salton Sea, and the inability of birds to adapt.

California has entered a period of drought not seen since the 1500s, says the paleoclimatologist Lynn Ingram of the University of California, Berkeley. Ancient stumps at the bottom of dry lakes tell the story of severe and vacillating drought across the Southwest from 800 to 1600 AD. Ancestral Puebloans abandoned the intricate cliff dwellings of Mesa Verde in Colorado and Canyon de Chelly in New Mexico around 1250 AD; the Mayan civilizations with advanced agricultural canals and dykes ended before 1000 AD. Now the West is entering another waterless era.

California is the most remarkable state in the Union. While every state is unique, California is a great civilization all on its own, geographically, culturally, technologically, and agriculturally. More than thirty-eight million people call the state their home—more than the entire population of Canada. It grows half the fruits, vegetables, and nuts we eat, produces most of our movies and TV shows, and boasts nine of our fifty-nine national parks within its boundaries.

From the deserts of Sonora, Mojave, and Joshua Tree to the redwoods and ancient sequoias of the north; from the coasts of Laguna, Malibu, and Santa Barbara to Monterey Bay, Mendocino, and San Francisco, from the granite mountains of Yosemite to the snow-covered volcanoes Mammoth and Shasta, and from the fields of strawberries and almond trees to the vineyards of Napa and Sonoma, there is no place in America, if not the world, that is so diverse. And risky—one has to be a gambler to live where fire, drought, mudslides, and earthquakes occur with regularity.

In the 1970s San Francisco's Hyatt Regency had a revolving restaurant at the top of the hotel with 360-degree views of the city and its sweeping bay. I had a day off from the TV movie *Death Be Not Proud*, John Gunther's moving story of his son. My husband, my son Jace, and I were having lunch there. The tower restaurant revolved on a huge disc like a giant 33⅓ LP record, giving us slow but continuous views to the horizon on this bright afternoon. We hadn't touched our dessert when there was a rumble, a bit of a shake, and the grinding squeal of a hidden winch as it attempted to keep us turning. Californians take their earthquakes, the small ones anyway, in stride, so there wasn't much of a fuss until the whole turntable began to wobble as it tried to right itself, the water in glasses sloshed over their lips, and the screeching of the wheel sounded like a DJ gone mad in a scratch studio. We were all evacuated safely. People weren't so fortunate in 1906 when a massive earthquake leveled the city, killing three thousand.

Another movie found me in a Hollywood motel in 1971 when the 6.5–6.7 San Fernando quake destroyed many buildings in the valley and took the lives of sixty-five people. I was awakened at 6 a.m. by a vase falling from a shelf and the continuous roll of the room. If I was to die I did not want to be alone, so I ventured outside to the pool, where the pool man and I watched the water careen back and forth like a teeter-totter, soaking the shrubs as it left the edges each time. The pool man never uttered a word and seemed more concerned about cracks in the cement than he was about his life, so I relaxed and began to experience the marvel of the Earth in motion. The birds that usually greeted the morning with song were nowhere to be seen, and I wondered where they rode out the aftershocks, which continued for the next thirty-six hours.

A massive earthquake is overdue on several faults, California is running out of water, and the temperature is rising. Its aquifers are being drained by indiscriminate wells, its mountains are losing snowpack, and fires will increase, straining the capacity to smother them. These might be insurmountable problems in any time prior to ours, but technology and new energy systems make it possible to mitigate damage and adapt to new conditions. Reduction of water use, new regulations on wells, and desalinization plants to pipe water from the ocean are all mitigation techniques that are workable with a commitment to protect wildlife and the environment.

California has never lacked for innovative thinkers, from technology's Steve Jobs to the champion of wilderness John Muir. When William Mulholland made Los Angeles what it is today by engineering the transfer of water through 233 miles of aqueduct from the Owens Valley in the eastern Sierra to the "city of angels" in 1913, the population jumped from 100,000 to almost 600,000 in a few years. Farmers planted almonds and lettuce with the irrigation of the Sacramento and Central valleys and began to feed the nation.

An era of dam building was ushered in to hold water and agencies were created to regulate its release. The first national debate on the environment began when John Muir, founder of the Sierra Club, voiced opposition to the Hetch Hetchy dam being built in Yosemite Park for San Francisco's burgeoning water needs. The citizens of the city, who billed themselves as "conservationists," declared that parks and their resources must serve the needs of *people* as well as nature. John Muir, a "preservationist," lost the battle in 1912, crying, "Dam Hetch Hetchy! As well dam for water-tanks, the people's cathedrals and churches, for no holier temple [than Yosemite] has ever been consecrated by the heart of man."

David Brower was born that same year and took up the preservationist's damnation of dams throughout his life. He too lost. The Bureau of Reclamation today maintains 475 dams and 348 reservoirs in the USA, supplying water to 30 million people and 1 in 5 western farms, which grow 60 percent of our vegetables. In addition, the hydropower serves three and a half million people. Dams have made life possible in the western desert.

Little care was given, however, to the complex systems of rivers and the wildlife within them. We made a mess of it. Most salmon in California, Oregon, and Washington will never recover from the blockage of spawning sites, the siltation from logging, and the pollution from development. The domino effect of this apex species' decline on other species continues. When salmon go upriver to die, the nutrients from their carcasses keep entire ecosystems alive, from tiny aquatic insects to Brown Bears. The only healthy wild salmon populations today are in Alaska, where Bald Eagles can be seen lining the banks during the annual fish migration in an extraordinary gastronomic orgy.

Brower may have lost the dam war, but he was instrumental in getting the Wilderness Act passed in 1964 designating certain areas where "man is a visitor who does not remain," the preservationist's dream. He is exquisitely profiled in one of my favorite books, *Encounters with the Archdruid,* by John McPhee.

The Wilderness Act was needed to mitigate the impact of human beings on wildlife—animals and plants invariably lose when people are allowed to occupy and exploit the natural resources of a protected area—but Congress can still seize water, oil, gas, logging, and mineral rights with the stroke of a pen. The environmental debate remains more charged than ever as the population increases and undeveloped land is appropriated. While more than half the globe is sparsely populated only 7 percent is protected. The war for pure wilderness is never over. The fight is worth it.

I was driving a lonely stretch of the Pacific Coast Highway one foggy morning from Los Angeles to Santa Barbara when hovering over the Ventura Mountains was a soaring giant, a huge black ghost

Condor, tagged #56, over the California coast

of a bird riding the thermals. I thought it was a small plane, it was so big. The California Condor's comeback is a thrilling success story. The largest bird in North America, with a wingspan of ten feet, it can live for sixty years and mates for life. In 1987 the U.S. government rounded up the last twenty-two in existence and captive-bred them in California. Today there are more than two hundred flying free in several western states and another two hundred still in captivity.

Lead shot in the carcass of deer and other wounded animals was continuing to poison these rare Condors. California led the nation with its ban on lead shot in 2014, taking on hunting groups, which had approved it only for waterfowl and did not want to switch to steel. California also passed legislation to protect Cougars from hunters and sharks from finning.

All roads lead to California in this twenty-first century. The state already generates about a third of its energy from wind and water and is jacking up solar. Along with her sister coastal states of Oregon and Washington, and Canada's British Columbia, California is formulating systems for mitigation and adaptation to drought, rising tides, and expanding populations. British Columbia claims it has already made a profit from new energy alternatives. These coastal leaders are forward-thinking men and women who see solutions for both people and the environment. If the funds are forthcoming from federal, state, and local governments, anything is possible.

Well, *almost* anything. We can play God only to a certain point. The little Spadefoot Toad took many millions of years to evolve into a successful desert dweller. We have just decades to implement the technology needed to rein in carbon emissions, mitigate drought and rising tides, and adapt to new living conditions. We can reduce carbon emissions, we can harness the sun to make us carbon free, but the huge weather systems that affect us are outside our control. They are global and spatial. El Niño, the jet stream, and sunspots may be predictable, but we cannot alter them . . . yet. I'm sure someone in California is working on it.

19

Killers

The Merlin shot out of the forest and nabbed the young Tree Swallow out of the air so quickly that all witnesses were caught off guard, including the chick's parents and me. As the Falcon made for the woods, the fledgling tight in its talons, the adult swallows took after him, joined by two others. It didn't matter; they would never catch up to one of the fastest birds in the world.

The Merlin pair was decimating the local population of Swallow and Warbler young. I ran along the trail weaving through spruces hoping to catch sight of the two Merlin nestlings as the Swallow was deposited before them. I was too late. The parents wheeled anxiously around the nest tree, screaming at my intrusion, and the chicks hunkered down out of sight.

Raptors such as Merlins are top carnivores in the bird world, just as canids and felines are among mammals—who would eat a Merlin if they could catch one.

Feral cats were wiping out ground-nesting birds in our woods one summer and grabbing those under the feeders as well. I counted three different feral cats over the course of the year. When a particularly large black one with fangs for incisors crossed the dirt road with four kittens in tow, I'd had enough. I pondered ways to do them all in. I prayed for Great Horned Owls and Coyotes to visit.

My mother used to say, "Beware of what you want, because you will probably get it." Within forty-eight hours a pack of Coyotes woke me at midnight baying and yelping like hounds from hell. Across the pond they sounded like twenty animals but were probably about ten frantically barking juveniles summoned by the alpha female to the kill. Then there was dead silence.

Was it one of the big Snowshoe Hares that were so tame around us? Or the Ruffed Grouse and her chicks in the sphagnum moss? Or the Muskrat in her mounded den in the cattails?

The rampage lasted for hours. In the days that followed it became clear that the feral cats had disappeared, along with Hares, Otter pups, Muskrats, and just about anything else that couldn't outrun or outwit the predators. There is something primeval, even frightening, in the howl of a pack as they come through the land.

I never feared Coyotes until the folksinger Taylor Mitchell was tragically killed by two of them while walking a trail in eastern Nova Scotia in 2009. It is highly unusual for Coyotes to attack human beings. I have spied them from California to New York and never gave them a thought except for the safety of my little dog Drama. Rangers soon shot the killer Coyotes, a third larger in Nova Scotia than those in the American West, but not part wolf as some believed. Why they attacked the young woman is unclear, but the Coyote is fairly new to the province, and in the wilderness of Cape Breton it may have developed no fear of human beings. Nova Scotia is such a benign place with regard to wildlife that it is a shock to think about Coyotes attacking. There are no poisonous snakes, or insects except for ticks, and no large mammals to fear except the occasional rogue bear if we happen to both be out getting blueberries in the same spot.

Although the Coyote is an effective predator, I know of no studies on the number of mammals and birds they kill annually. It is different with cats. A recent study by Smithsonian scientists surprised everyone when it reported that free-ranging house cats, strays, and feral cats kill a whopping 1.4 to 3.7 *billion* birds and as many as 20 billion small mammals a year in North America.

Cats are superbly designed for attack and can each take two hundred creatures a year. They don't always eat them either, preferring sometimes to toy with them, as any house cat owner can tell you. The American Bird Conservancy's "Cats Indoors" campaign has been urging the general public to keep cats inside, where they are protected from external diseases and not harming millions of birds. There are more cat owners today than at any time in our history and more people releasing them to the wild when they cannot care for them. While I would never endorse the annual trek to the river that my mother took with us children, cloth sack in hand full of newborn kittens to be drowned, it is irresponsible for cat owners to let their cats procreate with abandon or dump them in the woods as a problem for others. Neutering and spaying of feral cats, while well intentioned, does absolutely nothing to save the bird population and has not made much of a dent in the cat population, either; with as many as eighty million still roaming around, they can't all be captured.

Cats are the number-one killer of birds, after habitat loss, with windows coming in third. Almost a billion birds are estimated to die in collisions with glass where they cannot distinguish between reflection and reality. It is a sobering sight to see dead birds like the Northern Shrike, clearly not city dwellers, at the base of skyscrapers in Chicago during migration, or in Toronto, where massive new high-rises have especially egregious glass panes. Architects and builders are just beginning to be aware of the problem, and although it costs more, some are installing bird-friendly glass and designing with avian species in mind. Homeowners can prevent collisions by pasting raptor-shaped decals on large windows or placing shades strategically.

Wind power is the fastest growing energy system in North America, but it can be deadly to birds when huge numbers of the turbines are erected in migratory paths. Curiously, it is bats that are killed most often. One would think with their superb echolocation they would automatically avoid the gigantic blades, but mysteriously they seem to be attracted to them. The sudden drop in air pressure close to the blades is enough to burst their lungs and they fall dead to the ground. In migration from Canada to Central America hundreds of dead bats can be found beneath the wind towers. However, when the blades revolve just a fraction more slowly the bats survive. Wind farms are starting to absorb the 1 percent production loss to keep bats alive.

On a sylvan summer day in the mid-seventies I was driving a New York state highway in my little VW Rabbit with the top down, and ten-year-old Jon in the passenger seat, when a doe came out of the blue and leapt over the car, almost landing on Jon's head. She was killed instantly, but the fawn in her womb struggled to survive, its tiny hooves beating against her dark chamber. I would have slit open her belly and birthed it had I had a knife. We watched helplessly for ten minutes before life was extinguished.

More than two hundred people in the United States die annually in car accidents involving deer, and many more of the animals succumb. Although attacks from bears, Cougars, snakes, and sharks are on the rise as our population expands into the animals' habitats, the likelihood of death from these attacks is remote, less than a handful for each species every year. General car accidents in 2015 accounted for more than thirty-eight thousand people dying, while the murder of human beings by other human beings amounted to more than fourteen thousand. Human beings kill more of everything.

We are all killers. We may not acknowledge it, but it is not just the rapacious carnivore who kills. None of us gets a free pass. Every living thing on earth consumes some other living thing. There is even a fungus that grows only a millimeter every one hundred years, slowly digesting the rock it lives on . . . very slowly. Even rocks are living things if you believe, as Ovid did, that life is matter constantly undergoing metamorphoses—rocks break down over millions of years. We are all changing the equation of the earth with every footprint and every meal.

The glory of life is the incalculable variety of forms and functions that exist. No one knows how many species there are, but estimates put the number close to 8 million. Microbiologists point out, however, that in a single teaspoon of soil there can be as many as ten thousand bacteria. Only 1.5 million species of flora and fauna have been named and cataloged. The backlog for taxonomists is significant, and scientists keep reporting fifteen thousand new species annually. Most of the megafauna—the large animals over one hundred pounds, which include deer and human beings—have been cataloged and are the ones in danger of going extinct. It is no surprise the animals are dying; humans have been hunting them for 200,000 years.

After the mass extinction of the dinosaurs sixty-six million years ago, a small, sweet-looking weaselly creature named *Protungulatum donnae,* our ancestor, evolved. Paleontologists conjecture that it ate insects, was arboreal, and had a furry beige coat and a white belly. (A white belly? How do they know?)

I relate to this mammal: two eyes, a nose, padded paws with five fingers, live birth, and so on. We humans actually share 95 percent of her DNA—but then, we share 75 percent with the nematode worm and 50 percent with the garden pea, so that doesn't mean much, except that the template for all life is the same and only cements our connectedness.

In the ensuing sixty-five million years, the continents shifted, these little prosimian creatures evolved and radiated, and by the Eocene period, fifty-five million years ago, many survived on the isolated island of Madagascar to become lemurs or lorises in Southeast Asia. Elsewhere, around thirty-five million years ago, little monkeys appeared with larger brains and more forward-facing eyes. It is an exceedingly long time from thirty-five million years ago to around fifty thousand years ago, when "modern" human beings made their way out of Africa and began to disperse. Waves of hominids came and went as our ancestor was refined, creating complex tools, language, and art.

We were omnivores from the start, the more carnivorous among us living nearer the poles, subsisting on seal, other mammals, birds, and fish. Around twelve thousand years ago, some of us stopped wandering, tilled the soil, and planted seeds. A few animals were

domesticated. Clearing the land also meant killing wild animals: deer
and Wild Boars for food and wolves and Cougars out of fear. Food
and fear are the two great motivating forces of human behavior;
without either we would simply not survive. The rampant exploita-
tion of animal skins became possible with guns. Without needing
to meet the eye of our prey, we created another step removing us
from our familial connection to animals. It is easier to kill when the
"otherness" of your victim is emphasized, when he is an object, not
a living being like yourself.

The wild animal fear factor is primal, and a healthy respect for
them is necessary, but killing them as "nuisance" animals is war-
ranted only when they are diseased or documented man-killers.
Wildlife Services is a federal program of the U.S. Department of
Agriculture. It was called Animal Damage Control for more than a
hundred years, initially providing rodent and pest control for crops.
It has developed into a well-oiled killing machine for just about anyone who wishes the removal—death or otherwise—of an offending
animal, and has the funds to pay for it, at least partially. The rest
is borne by taxpayers, who have no idea they are bankrolling the
deaths of as many as four million birds and other animals annually.

Twenty-two thousand and five hundred Beavers were killed by
Wildlife Services in 2014 alone.

The Beaver is one of nature's great creatures, clearing land inde-
fatigably to build dams to house its young, safe from predators.
Streams and lakes back up and create spawning areas for fish and
amphibians, attracting birds and mammals of all kinds.

Beavers can certainly be annoying to human beings. In our case, a
Beaver decided that the sluice of our pond, which drained overflow
into the ocean, was a perfect spot to build a dam. Had he succeeded,
the water would have backed up, eventually flooding the first floor
of our house. The Beaver spent every night cutting saplings and strategically placing them over the sluice. Every morning friends and
I would haul the night's construction away. This did not deter the
master builder. The game of patience became intense, neither of us
giving up. Finally, after many weeks, he was sufficiently discouraged

to move on to another pond. My admiration for his tenacity was immense. There is absolutely no good reason to annihilate 22,500 of these magnificent animals.

Wildlife Services is at work eliminating most of the existing Mute Swans on the East Coast—these are the beautiful big white birds of picture postcards and swan boats. The bird is designated an "invasive species." It "invaded"—that is, was brought in from Europe for its beauty—more than a hundred years ago, which trumps its other designation of "protected species." So much for being special. In the Chesapeake Bay the voracious appetite of these swans resulted in the decline of the aquatic vegetation that shelters the crayfish, spawn, and crabs on which the economy depends. What is often not mentioned is that the real culprit is polluted runoff from pesticide- and herbicide-soaked lawns and farms that line the shores of the bay, putrefying the vegetation. People first.

Wildlife Services does no service to wildlife at all; it is in the pocket of people who have no respect, much less patience, for wildlife. The list of victims in 2014 is the size of a small book and includes 454 River Otters, 250 Barn Owls, 5,000 vultures, and 8,971 ravens, one of the most remarkable and intelligent birds in the world.

There are always more humane solutions to be had. The irruption of Snowy Owls from Canada in the winter of 2013, when they could be found from Maine to Idaho, caused problems on the open terrain of airport runways, which mimicked their tundra home. New York killed three of the Snowy Owls and was about to slaughter a dozen more at JFK airport when the New York Audubon Society asked its members to get on the phone. We bird lovers knew of a successful program at Boston's Logan Airport begun decades ago by one thoughtful man, Norman Smith, who personally wrangled each and every Snowy Owl and released them in safe territory. In 2014 he safely released 120 of the runway owls. Audubon was successful in halting the New York owl slaughter in a similar way. That is, until 2016, when a court ruled that airports had a right to kill them.

Four million animal deaths at the hands of government programs may pale in comparison to the number of birds alone killed by cats and windows, but as taxpayers we have their blood on all our hands.

Wildlife Services, hunting, cats, and windows take an enormous toll on the lives of birds and other animals. But it is loss of habitat that is the primary and most insidious killer of wildlife. The human population keeps expanding despite predictions only forty years ago that it would level out when women of the world had enough education and economic security to take control of their bodies. It has not worked out that way. The First World countries of Europe and North America have reduced their populations, and some populations are even in decline, although with the infusion of immigrants from other parts of the world the numbers will rise. But women are still repressed the world over, refused education, enslaved, and provided little health care. Until the political and spiritual lives in these repressive countries change, the human population will continue to grow, and the animal population will decline.

When we first moved to Nova Scotia in 1998, I would sometimes forget the speed limit was in kilometers per hour, not miles per hour. I have always owned a convertible, loving the wind on my face and hair and the sight of a full arching sky where raptors circle and clouds pillow on the horizon. I joyously gunned my Audi convertible one day on a lonely stretch of Canadian highway until the speedometer registered close to 100. The speed was exhilarating. Only later did I realize that speed sign meant 100 kilometers per hour, not the 160 kilometers I was driving, and how lucky I was not to have been arrested by a Mountie.

Cars are amazing inventions. I would be bereft without mine. I understand why everyone in the world wants one. China has created a huge middle class in the past decade with wealth enough to buy not only ivory but also cars. The number of cars in the cities of China will probably pass the 189 million in the United States by 2020. The infrastructure is not in place to handle the number of roads necessary or the carbon emissions from exhaust pipes. One Chinese traffic jam in 2014 extended for several days and more than sixty miles. Pollution in Beijing has become so deadly that on the U.S. embassy rooftop it has measured 886 parts per million, more than twice the most extreme measurement in the United States and considered highly toxic. Yet the population keeps growing and solar-powered vehicles are still in the design stage.

Stephen Hawking has a sobering perspective: "The human race is just a chemical scum on a moderate-sized planet orbiting around a very average star in the outer suburb of one among a hundred billion galaxies." We may never know if we are alone in this vast universe, or whether we are superior or inferior to creatures on other planets in other solar systems, but we do know where we rank in the hierarchy of our own planet at this point in time. We are dominating all existent life-forms. As the most adaptable mammal in the world we could in theory feed and house all of us satisfactorily if we had the will to make it happen. We also could coexist with our fellow creatures if we had the will. But we have hurled ourselves headlong into destructive practices, heedless of the consequences to Mother Earth and to ourselves. Stephen Hawking says everything could perish in the twenty-second century.

It was springtime in the woods of northern Pennsylvania, the expansive and deep forests that go on for hundreds of miles. We looked down on the "little grand canyon" of the state, a lovely glacial ravine with excellent trout fishing in the Pine Creek–Susquehanna River, which meets up with the Allegheny and Genesee to form the continental triple divide. Each river takes a different course, the Allegheny heading ultimately to the Gulf of Mexico, the Genesee to the North Atlantic, and the Susquehanna to the Chesapeake Bay.

My birding friends from Pennsylvania Audubon and I were listening to courting warblers in the oaks and pines when we finally saw the fire-engine red of the Scarlet Tanager as it skirted the treetops, burbling its lazy robin-like song. We have only four bright red birds in North America: the Northern Cardinal, the Vermilion Fly-catcher, the Summer Tanager, and the Scarlet Tanager. They dazzle the eye in their breeding plumage.

Male Scarlets arrive in the forest to scout nesting sites in May, singing from a spot high in the canopy. Although they are a common bird of the eastern United States they can be difficult to see so high up. Tanagers need large tracts of forest; they are secretive and try to hide from Brown-headed Cowbirds, which parasitize other birds' nests by placing their eggs in them. But Cowbirds rarely venture deep into the forest, preferring to do their skulduggery near forest edges where they can see the comings and goings of different birds and then dash in to make a deposit. Pennsylvania's extensive oak forests are coveted habitat for Scarlet Tanagers, and it is estimated that 10 to 13 percent of the entire population nests there. The Scarlet Tanager needs four to eight acres of unfragmented forest for breeding.

In a twist of fate, 1.5 million acres of the glorious Pennsylvania forest hosts the lucrative Marcellus Shale beneath it, the pot of gold for natural gas extraction, now expedited by hydraulic fracturing, or "fracking." The Department of Conservation and Natural Resources is charged with managing "the state forest system for many uses and values—including natural gas development—all the while protecting its ecological integrity and wild character," says the state website. Tell that to the Tanagers.

Pennsylvania has a long history of drilling. It took advantage of massive oil reserves throughout the twentieth century. Hydraulic fracturing for natural gas seemed a natural and exciting new development for the state and an answer to economic depression. By 2015 the state had more than six thousand active fracking wells and was planning to increase that tenfold.

I wanted to fly over Tioga County to see what fragmentation of the forest was occurring with the well pads. Anyone who has ever flown over a mine, logging, or drill site has seen the wounds made by extraction. Even the cleanup later, if there is any, reveals a pale place with little relationship to what once was. Half a mountaintop blasted away to reach veins of coal is covered over with grasses and lines of trees that rarely entice original inhabitants back in the next twenty or a hundred years. The complex relationship of the soil to the plants and insects that are born from it is compromised forever.

Extraction always takes its toll on the body of the earth, as it does on the body of a human being. The extractive practices of timbering, mining, and drilling are surgical operations: remove the wart, the tumor, or the organ and sew it back up. Fracking is like injecting gallons of water and chemicals into a main artery to explode the blood vessels and suck out the blood. There is hardly a corpus left after the operation. The damage inside is extensive.

Our single-engine plane began its journey over the pristine forest, which covers almost 60 percent of the state, sheltering the great rivers, which cut into the bedrock lining the steep mountainsides. There is wildness to it, a comfort in the belief that wild things are safe even though you cannot see them. Black Bears, White-tailed Deer, Opossums, Porcupines, and other common denizens of the northeastern United States are plentiful and have had little to fear in the past except hunting season.

As we veered south, the well pads for the drills and the cleared ground that each pad needed came into view, neatly cut out of the forest, or nestled beside farms like blocks on a chessboard. Every well site, or "pad," for fracturing needs three to seven acres, leveled so trucks and drills have open access. There may be as many as 150 contiguous well pads covering 1,000 acres, with numerous compressor stations in between. The drill is bored into the ground vertically for a mile under the enormous pressure of a million gallons of water and chemicals. It then bores horizontally through the shale, fracturing it to extract the gas. What is not seen from the air is the destruction below the earth, the crisscross of pipes and the blasted shale rock extending for miles underground.

The natural settling of the earth's mantle is compromised by fracking. More than a thousand earthquakes occurred in 2014 in Oklahoma and Texas alone because of the lubrication of fault lines due to fracking. The noise from constant drilling for weeks on end at a well pad is enough to drive most mammals and birds from the area,

*Imagine this fracking site multiplied by 1,000 or 100,000 contiguously
and you understand what a fragmented forest is.*

and the seismic activity underground must displace as many creatures below. Trucks by the hundreds rumble through the roads day
and night delivering equipment and removing waste while the stacks
of fire burning off methane light up the night sky. Local streams,
lakes, and rivers are often used to cart or pipe the million or more
gallons needed for each well, depleting natural flow and water levels.
Wendell Berry's environmental golden rule, "Do unto those downstream as you would have those upstream do unto you," holds no
sway with frackers.

I could not imagine the Tioga County landscape with the 60,000
to 100,000 well pads planned by 2030. The local people put up with
the stygian scene because many get a healthy check for leasing their
land. The sensitive Scarlet Tanager doesn't have a say in the matter.
People say it won't go on forever. But will it? The toxicity of the
chemicals used may be with us for many generations.

Fracking took off in 2007 when an exemption—not a
coincidence—to the Clean Water Act called the Halliburton (or
Cheney) Loophole was passed by Congress giving companies the

right to withhold information about injected chemicals. The Environmental Protection Agency's 2004 report stated that "the injection of hydraulic fracturing fluids into . . . wells poses little or no threat to USDWs [underground sources of drinking water] and does not justify additional study at this time."

More than six hundred different chemicals have been tried in fracking fluid, including concentrations and compounds of chloride, bromide, strontium, barium, benzene, methane, and the NORMs (naturally occurring radioactive materials—uranium, radium, and radon). These chemicals have all been suspected in the injection of wells and have been found in the "flowback," the fluids that return to the surface of the wellhead during fracking. The chemicals also live on in the slurry of water and shale after the gas is siphoned out, and in the air. They infiltrate streams and drinking water to the point where methane in tap water can be lit with a match.

The chemicals are toxic to animals and plants. People living near well pads have reported an increase in nosebleeds, abdominal pains, headaches, rashes, and diarrhea. Scientists have found a correlation between a rise in hospital visits and that of fracking sites in three counties in northeastern Pennsylvania. The number of chemicals impacting human and animal health increases with each new study. These chemicals can have very long lives, especially the radioactive ones.

My Nebraska grandfather, Dr. Daniel Quigley, went to visit Madame Marie Curie in France in 1913. He returned to Omaha with two small chunks of radium, and opened the first Radium Hospital west of the Mississippi to treat cancer. My dad told me that the radium was kept in the icebox in their kitchen. Its volatility was known, but not the creeping deadliness of its poison. My grandfather lost his middle finger—the one he used to place radium on cervical cancers—and his skin was the color of a bad sunburn for the rest of his life. He almost glowed in the dark. My grandmother developed a tumor and half her brain was removed. She never functioned fully again. The road to hell is paved with good intentions.

As our little plane circled back toward the airport and the view of well pads spotted the landscape to the horizon, Tioga County's planning director, Jim Weaver, in the seat behind me, said through the earphones, "You are looking at the last gasp of fossil fuels."

I believe him. Our reliance on fossil fuels is coming to an end. Economically it will not make sense much longer, when the cost of holding down carbon emissions exceeds the costs of extraction. Yet as it sputters out we continue to inflict damage on our earth, air, and water, without calculating the cost to human and environmental health. Our country is at last independent of foreign oil and gas and is riding the wave of prosperity and cheap prices at the pump like a rodeo cowboy on a Brahma bull holding on for as long as he can, heedless of the crash to come.

It was a cool February day. Something large caught my eye above the ball field of my granddaughter's school. Two large birds seemed to be in some form of deadly combat in the sky. As they plummeted toward the ground I saw they were two Bald Eagles, their talons locked in ritual courtship behavior as they spiraled around and around together like a gyroscope before landing lightly on the ground.

This was one of the more remarkable sights in my lifetime of bird watching, made all the more remarkable because the Bald Eagle was almost gone in my youth and made an astonishing comeback when the chemical DDT was banned and the bird's eggs were finally viable again. We human beings made the difference; we turned things around. I feel proud and as deeply grateful as I do when human lives are saved through an act of courage.

Pope Francis, in his eloquent 2015 encyclical on climate, wrote extensively about the connectedness of all living things and our interdependence. The web of life is so complex that the whole is not possible to grasp and we must take it on faith that even the lowliest of us, the insects, microbes, and fungi of the world, have a place and a purpose. This is as much a secular and scientific argument as it is religious. There are things we must keep sacred.

When Rachel Carson's own encyclical on deadly consequences, *Silent Spring,* was published in 1962, calling for an end to DDT and the spraying of other chemicals that infiltrated our water systems and our bodies, there was hope that this was an end to these practices, that the lesson was learned. It was not. DDT was simply repackaged for poor countries abroad and chemists came up with new ones for us to use here at home. More than fifty years later we are more reliant than ever on chemicals to kill pests, create new fabrics, and "protect" us from cradle to grave.

This is the century when we need to learn better ways to heat our homes, light our buildings, and keep our water pure in the face of enormous climate challenges, but we continue to infuse seeds, food, waterways, and the atmosphere with "quick fixes," polluting the very life forces on which all living things depend. The incidence of disease is increasing and new undiagnosed ailments are everywhere.

Dr. Frederica Perera, a molecular epidemiologist with the Columbia Center for Children's Environmental Health, has been doing long-term studies on young people from infancy to adolescence. She has found significant changes within the molecular structure of their bodies due to chemical exposure. Our homes are inundated with chemicals in everything from flooring to fabrics, soaps to salves. The endocrine system can take only so much before it begins to make changes, either destroying tissue or altering it in an effort to thrive.

Some creatures cannot take the assault at all. Pesticides such as neonicotinoids, used mainly on crops, cause disorientation in the homing systems of many insects, including bees. Neonics began to be used widely in 1996; two neurotoxins especially, imidacloprid and clothianidin, proved to be highly effective killers of just about any bug you can name: aphids, Cane Beetles, cockroaches, carpenter ants, termites, fleas, Japanese Beetles, thrips, and locusts, to name some of the undesirables. If these hardy insects can be leveled by the neonics, what do these poisons do to the legions of beetles that spade the earth, to the wasps and butterflies that pollinate the flowers, to the spiders that catch flies, and to all the microscopic creatures that E. O. Wilson calls "the little things that run the world"? Manufacturers claimed that neonics were less toxic to birds and mammals. How can that be when many birds and small mammals eat insects, even dying ones?

A huge eighty-year-old hickory tree that dominated our Putnam County lawn was blighted by Gypsy Moths one summer, killing the leaves and reducing the nut crop to nothing. A reputable tree company promised me that spraying the tree would not affect the fish in the lily pond beneath it or the animals that came to drink from it. They lied. All the Goldfish and Bullfrogs went belly up within hours, dragonfly bodies draped the lily pads, while earthworms desiccated on the stone walls. I never used a pesticide again, or a herbicide on my lawn, flowers, or vegetables.

Almost all corn crops today are treated with neonics, as are soybeans, potatoes, leafy greens, nuts, beets, and most fruit trees. The list is long. Trees are treated to prevent invaders like the Emerald Ash Borer and Wooly Adelgid; within sixty days the entire tree is infused with the neurotoxin from the trunk to the tip of the tallest twig, and the insects living on the host tree cannot thrive. What happens to birds like woodpeckers and warblers, which comb the bark and leaves of these trees for insects? Do they ingest the poisoned ones? Or do they starve because there are no more insects? Has anyone done tests on these insectivores that may take hundreds of insects a day, especially when feeding their young? The answer is no—not to date, anyway. It was enough for scientists at Bayer AG and the FDA to green-light neonicotinoids when tests showed that mammals were not susceptible to low doses of the pesticide—high doses such as a spill, yes. Well, I guess spills never happen, right? Sound familiar?

By 2006 Honeybees were suffering what came to be known as colony collapse disorder. While all problems for bees may not be attributable to neonicotinoids, clothianidin in particular has been shown to interfere with their ability to return to the hive and for the queen to thrive, a kind of nerve gas for insects.

Bayer AG's profits are a billion dollars annually on the sale of neonics, the most-used pesticides in the world today. Bayer may have convinced the U.S. Food and Drug Administration that their science was sound, and government regulators may have bought into it, but the science did not take into account the cumulative effect of many years of ingestion by our littlest creatures. It has taken less than twenty years for the most wondrous pollinators on earth, our bees, to begin to suffer.

The birds declining most rapidly today are the insectivores— swallows, swifts, nightjars, flycatchers, small raptors, warblers, thrushes, and many more. But almost all birds will eat insects some of the time. When neonicotinoids get into water in runoff from fields, the pesticides are highly toxic to aquatic invertebrates, on which many fish and some birds feed.

Scientists in the Netherlands became deeply concerned with the high neonicotinoid concentrations in insectivorous birds, estimating a decline of 3.5 percent annually, hardly sustainable in the long run. In 2014 they pointed to "potential cascading effects of neonicotinoids on ecosystems."

A cascading effect, or trophic cascade, is when the entire infrastructure of an ecosystem collapses like dominoes. It often begins from the top down when an apex predator is no longer around. Or, in the case of pesticides, it can happen from the bottom up. The neonics leave residue on flowering crops or garden flowers, which is then ingested by bees, wasps, or aphids, which are in turn eaten by insectivorous birds, large insects, or small mammals, which are in turn eaten by larger scavengers or carnivores. Or the neonicotinoids kill the insect populations on the crops or garden flowers or tree leaves, leading again to the poisoning or starvation of insectivorous creatures. The loss of these birds and insects means the loss of food sources for larger birds and animals higher up the food chain until at the top the predators are left with fewer prey and the system approaches collapse.

"Future historians may well be amazed by our distorted sense of proportion. How could intelligent beings seek to control a few unwanted species by a method that contaminated the entire environment and brought the threat of disease and death even to their own kind?" Rachel Carson wrote this more than fifty years ago. Not much has changed in our attitudes, and the situation has gotten worse. The *Silent Spring* of our generation is the insidious and deadly game of harmful chemicals in almost every arena of our lives: herbicides, pesticides, processed food, public water, synthetic fabrics, medicines, and energy sources.

Rachel Carson's dedication over fifty years ago changed the course of history in North America when DDT was banned. I gaze in awe today at the sight of a Bald Eagle or a Peregrine Falcon, giving thanks for this remarkable woman and the legislators who finally listened to her, after first repudiating her science as that of a hysterical woman. It is what gives me hope for the future. What also gives me hope are the many people who are growing and buying natural chemical-free food and products and trying to make them available to those from all walks of life. Farmers' markets have become weekend community gatherings in towns all across America.

You may have noticed a decline in the number of butterflies, fireflies, dragonflies, or swallows, and yes, even mosquitoes, where you live. Pay attention. Think poison.

20

Bhutan

Transiting from India to Bhutan makes clear what the weight of 1.2 billion people can do to the land. Our van plowed through the dusty, highly congested streets of the northern city of Assam, Guwahati, dodging bicycles, horse carts, pickup trucks, and sacred cows. In the hours it took to reach the border of southeastern Bhutan we birded from our windows and in the shimmering heat of the afternoon managed to see two huge Storks on roadside trees. They sat there in their white flowing feathers like prelates on a Sunday morning, oblivious to the melee beneath them and claiming ownership of the spot as they had for millennia. The Greater Adjutant Stork is an endangered species, its territories diminished through the years by the incessant influx of more and more people. Indians respect animals; they allow them free rein and let them live—if they can. This accounts for odd juxtapositions such as the Storks we witnessed that day, unfazed by the industry below. They hang on, barely.

But there are no vultures. India's vultures are almost completely gone, in the most rapid decline ever documented. In 1985 the White-rumped Vulture was considered perhaps the most abundant bird of prey in the world. Two decades later 99 percent of its population had collapsed, as had those of other vulture species in India. Only the collapse of the Passenger Pigeon, considered the most abundant species in North America in the late nineteenth century, is comparable.

In the case of the vultures, diclofenac, an anti-inflammatory drug commonly prescribed by veterinarians, is the main culprit. India, a Hindu country, reveres its cattle as the repository of former humans waiting to be reincarnated. They wander at will through the streets, the fields, and the alleyways. The old arthritic bovines responded so well to the injection of diclofenac to reduce pain that the practice became widespread. The animals also die in the open. The vultures are supremely effective at picking the carcasses clean within hours of death, thereby preventing the spread of disease. Unfortunately, the birds cannot tolerate the drug the cows have been given and the birds die—millions of them.

This presents a huge problem for India as free-ranging dogs take over the business of carcass disposal. There are eighteen million of these dogs in India; thirty thousand human beings die annually from rabies because of dog bites. The packs of dogs running through neighborhoods are frightening to behold as they seek food wherever they can get it.

For the Parsi, a small sect that consecrates its dead to the vultures on its "tower of silence," the absence of the scavengers has been disastrous. Vultures are the intermediaries between heaven and earth, and without them the bodies of the dead can languish in the open for as long as six months, disturbing the cycle as well as the peace of mind of Parsi relatives.

As soon as diclofenac was identified as the killer, India banned it in 2006, followed by Nepal and Pakistan. The alternative drug meloxicam does not seem to affect the vultures and has widely replaced diclofenac for veterinary use, although not for human use, so there are still some dying birds. India has spent billions of dollars on the problems resulting from the crash of the vulture population. The birds are worth their weight in gold, almost literally. The vultures will come back—nature is resilient if given half a chance—but it will take a long time.

Our Bhutanese team. Hishey Tsering, our guide, is on the right, 2012.

We passed into Samdrup Jongkhar, Bhutan's most southerly territory, and into the mythical Shangri-la of the world. The difference from India was readily apparent. Assam has 6 million people, and Bhutan has 800,000. There is room to breathe; foliage draped the wet embankments as we began to climb the winding roads, wide enough for only one vehicle.

Up and up we went, stopping for birds whenever a good spot allowed. Wildlife was abundant. In the first twenty-four hours we saw two large cat species, the Jungle Cat and the rarely seen Golden Cat. The latter, a quarter the size of a Cougar, dashed across the road at dusk, leapt up the embankment, and disappeared into the forest. Wild cats are rarely seen at all, and the Golden Cat is seen less than most. The "golden" in the cat was not visible to us in the low light—we saw only a sleek brown body—but the powerful grace with which it lifted off the roadside is the mark of all great cats and thrilled us with its wildness. Later, at dinner, when we congratulated ourselves on seeing this most elusive animal, our guide Hishey said that I made it possible. "How's that?" I said. "Remember, you asked for a pit stop a half hour before? We never would have crossed paths with the Golden Cat if we hadn't stopped earlier." Ah, we weren't in Kansas anymore.

And we were not. The thinking was different, and this became more apparent every day as we traveled from the very rural eastern part of Bhutan to the more developed west. Places were not far apart—Bhutan is only as big as Switzerland—but it takes time on the perilous mountain roads to get from one place to another. There was not a lot of traffic, but when there was, it was hair-raising. The vehicles coming in the opposite direction were usually big brightly decorated trucks with fanciful tiger faces painted on the grille; they belched acrid diesel fuel as they squealed to a stop on a turnout inches from the edge of a five-hundred-foot drop. These drivers were masterful, and ours, Pala, was the best driver I have ever known. He seemed to have a sixth sense and while driving managed to spot birds at the same time, shouting out where they could be seen.

The Bhutanese are kind people, and playful. One afternoon I spied Hishey on a forest road trying to catch leaves as they drifted to the ground, jumping in the air like a basketball star and then scooping a leaf into his arms. He was joyful, as were the other Bhutanese who accompanied us and whom we met on our way. He was also one of the brightest men I have met in my travels—intuitively bright, well educated, and knowledgeable about wildlife.

Bhutan is the last monarchy of the Himalayan countries. They have managed to repel forces from India, Tibet, and elsewhere repeatedly through the ages. The five "Dragon Kings" of the last one hundred years have ruled wisely, and the current king, Jigme Khesar Namgyel Wangchuk, has the status of a rock star among the people. They hang his portrait and that of his beautiful queen prominently in their homes, next to Guru Rinpoche, who brought Tibetan Buddhism to Bhutan in 747 AD.

They are a beautiful couple who came to the throne when Jigme Khesar was only twenty-six. His father, Jigme Singye, realizing that most of the population was under thirty years old, insisted on a transition to a modern democracy and abdicated in 2006 so his young son could assume power and spearhead the new electorate. Jigme Singye also introduced the concept of "Gross National Happiness" as a national policy, promoting the idea that development of a nation was not solely based on economics but on a number of factors including health, education, and community well-being. Government surveys regularly assess the happiness of their citizens through interviews and community meetings.

The current king and his queen, like the reigning monarchs of England, travel throughout the country meeting with Bhutanese everywhere and gauging how they are. In one inn where we stayed there was a little temple next door with a picture on the wall. Our innkeeper's husband was in a wheelchair as a result of a terrible car accident. In the picture King Khesar is touching him gently under the chin. "You have suffered much, being paralyzed so young," he said.

This compassion seems inherent in the general populace, not just the monarchy. My husband, who has some difficulty walking, was out at dawn one morning trying to keep up with us birders on a narrow path outside a small village. He urged us to go ahead and sat on a large rock to wait for our return an hour hence. In the houses nearby families were starting their day; children were washing their faces and brushing their teeth from cold-water buckets by the door, as woodsmoke unfurled from fire pits and breakfast grains bubbled in pots.

Bhutanese houses are fairly uniform in their architecture; the government prescribes a general style as part of its larger plan of one culture. It is a pleasing look. A symmetrical rectangle of whitewashed stone or pounded earth is fronted above by wood decorated in colorful patterns of animals and plants, and then roofed in a graceful slope. The cornice might have a bright red phallus dangling from it, as a symbol of fertility.

Ed took up his spot on the smooth rock a few hundred feet from some houses and began to meditate as the sun came over the mountain. We had not left him more than ten minutes when a man emerged from his home in the chill air concerned for Ed's well-

being. Ed assured him he was fine, whereupon the man invited him
to join his family for breakfast. And so Ed passed a delightful hour
drinking hot tea and eating biscuits while the rest of us combed the
surrounding hills with growling stomachs. He encountered more
acts of kindness in the weeks that followed.

Because hunting is uncommon and the human population is low,
one experiences Bhutan as a more balanced natural landscape than
elsewhere. Some mountains are off-limits to human beings entirely,
being the sacred domain of gods, as in neighboring Sikkim. This
includes the highest unclimbed mountain in the world, Gangkar
Puensam, at 24,735 feet. Tigers have been spotted at the extraordinary altitude of 15,000 feet, and while they are not plentiful anywhere Bhutan seems to have a few hidden in its forests.

Our group of twelve was here with George Archibald, cofounder
of the International Crane Foundation, primarily to see the Black-
necked Cranes on their wintering grounds in two valleys. We were
not doing any hard-core birding, descending into the brush to search
for a particular species, nor were we going off-road for mammals.
Our van stopped often, however, and we were rewarded with many
sightings of both.

I awakened each dawn snuggled under my comforter in the heat-
less rooms of monasteries and inns, to the song of the Blue Whistling
Thrush outside. This sweet songbird is as ubiquitous as our American Robin and as welcome. As the day commenced, birds came to
life everywhere in song and flight. The scenery was spectacular, from
sunrises that bathed the snowy mountaintops a baby pink to golden
fields of wheat studding the valleys and still threshed by hand.

In the Bumdeling Wildlife Sanctuary we saw our first Black-
necked Cranes, which had flown over the Himalayas to settle in the
harvested rice paddies for the winter, gleaning the leftover grain in
the runoff. Their numbers were down due to recent flooding of the
paddies in the river valley and dog incursion. Dogs were everywhere,
foraging in small packs and chasing whatever they could. They were
not vicious, probably because the Bhutanese respect all living things
and do not beat them. Puppies spilled out of baskets on village

streets, narrowly missing the wheels of a whizzing truck. The dogs procreate with abandon, and this causes problems for threatened species like cranes.

On the sand islands of the low river in Bumdeling we counted only seventeen of these magnificent birds, with their whitish-gray bodies and black necks, strutting the banks like stately priests on high holy days. They are almost five feet tall, and, as with all fifteen species of crane, they are one of the oldest orders of birds in the world, from the Eocene period, ten million years after the demise of the dinosaurs. If the Pterosaurs hadn't gone extinct I might have believed cranes were their heirs.

All crane species are in trouble due to human incursion, despite the fact that they are revered from Japan through India. They are found on the five continents of Asia, Africa, Australia, Europe, and North America. In the United States and Canada we have the Sand-hill Crane, which numbers in the hundreds of thousands, and the Whooping Crane, with less than four hundred in the wild, making it the rarest crane on earth. We do not revere the birds as Asians do. There is still hunting of Sandhill Cranes every autumn, with guns blasting them out of the sky as flocks migrate south for the winter.

George Archibald is one of the most famous ornithologists in the world. He helped bring the Whooping Crane back from sure extinction when there were only a few dozen left in the wild. As a student in the 1970s he obtained a young captive-raised whooper from Texas and named her Tex. He spent many days studying her in her enclosure, and before long this lanky six-foot-tall avian Texan fell in love with George, chasing away all humans who paid attention to him. Tex had no interest in any male crane and had eyes only for George, which presented a problem, since he was hoping to breed her. So George did the only thing a self-respecting young biologist could do: he danced with her, mirroring her excited courtship leaps and whirls, prancing around her until she was ready to be held and artificially inseminated. There is a charming video of George and Tex dancing. A captive-bred chick was born months later, named Gee Whiz, and he went on to father a number of chicks of his own.

George realized that imprinting on human beings was problematic and began using crane hand puppets to feed the new hatchlings and later human beings dressed as cranes as they matured. Today the International Crane Foundation in Baraboo, Wisconsin, continues breeding Whooping Cranes and teaches them to follow an ultralight plane costumed like a giant crane on their migration south, reintroducing them to ancestral homes in Florida and elsewhere. The ICF is the only place in the world where one can see all fifteen species of cranes.

In Asia the draining of marshes and wetlands for human development and agricultural purposes presses the cranes into smaller and smaller areas and their numbers decline. George travels much of the year to countries like Russia, Mongolia, China, South Korea, and Africa, meeting with heads of state, local leaders, villagers, and scientists to encourage everyone to protect these remarkable birds. He is one of the most gracious men I know, with an abiding love of his fellow human beings. He has taught me that the most important thing in conservation is inclusion, embracing one and all in the pro-

Cofounder of the International Crane Foundation, the ornithologist George Archibald has time for everyone. Here he talks with a young monk at a dzong, Bhutan, 2012.

tection of species, and letting them know about the valuable natural resources people have in their area so they can be proud and protect them.

Twice a year the Black-necked Cranes make the journey over the high Himalayas from Tibet to Bhutan and back. It is a primal event. Aldo Leopold put it this way, writing about Sandhill Cranes a half a world away:

> Out of some far recess of the sky the tinkling of little bells . . . then a clear blast . . . and finally a pandemonium of trumpets, rattles, croaks and cries . . . at last a glint of sun reveals the approach of a great echelon of birds as they sweep a final arc of sky and settle in clangorous descending spirals to their feeding grounds . . . such a place holds a paleontological patent of nobility.

Ed and I had accompanied George once on his annual pilgrimage to Nebraska, where every March 600,000 Sandhill Cranes and a few whoopers stop over on the sandy islands of the Platte River to fuel up before they head off for Canada. It is one of the great migrations on the planet and an astonishing sight to witness when the birds circle down by the thousands and pack in together for the night, as they have for millennia.

The Phobjikha Valley in central Bhutan has been the wintering spot of Black-necked Cranes for just as long. We watched as hundreds descended at dusk to their roosting spots near ponds on the valley floor. They are legally protected in Bhutan, China, and India. The Phobjikha residents watch over the cranes, keeping the waterways clear for them and chasing away dogs and foxes. The government of Bhutan has even agreed to change the power lines that string across the valley, sometimes catching the cranes by their necks, by burying the lines underground. Tourism is a growing industry, and the cranes are a real draw.

At 4:30 the next morning four of us walked the valley edges through frost-covered fields and over tiny sparkling streams, our way

Prayer flags wave in the wind as we wait in Phobjikha Valley
for the cranes to fly in for the night.

lit by a million stars above, brighter than I have ever seen. We were
intent on counting the roosting birds before they headed for their
feeding grounds at first light. George and a young Bhutanese fellow
from the local Royal Society for Protection of Nature did counts
through their scopes and a colleague and I tallied the numbers.

We waited on the chill grass for the birds to stir, and by 5:30 they
were lifting off in family groups of three and four, honking their
departure. My fingers didn't bend well in the cold, scratching the
pencil point across the notepad. George and Jigme did four counts
before they reached a consensus of three hundred adults and thirty-
eight chicks, a successful fledge rate. The cold was penetrating our
bones, so three of us headed back over the sunlit fields and streams,
passing grazing horses with their colts and a few cranes with *their*
gangling "colts," as they are called, picking at wheat chaff. George
stayed behind to watch and photograph his beloved cranes as he
always does, an acolyte at the altar of creation.

The bird I most wanted to see in the Himalayas, if not in the world,
was not a crane. It was a pheasant. These mountains are known for

Black-necked Cranes and their "colts"

spectacular pheasants, and to my mind the most spectacular of them all is the Monal Pheasant. I had seen the bird before in the captivity of the Bronx Zoo, where it made a well-worn path in its six-by-six-foot enclosure racing madly back and forth in front of a wall of wire. This was not the wild bird of my dreams, but it gave me an indication of the colors to come. I told Hishey of my fervent wish. He promised nothing.

Our mammal list grew daily with glimpses of Yellow-throated Martens, Assam Macaques, Gorals, Striped Squirrels, Capped Langurs, and Red Foxes. A troop of endangered Golden Langurs greeted us one afternoon as we alighted from the van for a walk through falling autumn leaves. The long blond tail of the primate gave him away as it hung straight down through the foliage. A mother and her baby, clothed in a mantle of softest palomino plush, leapt twenty feet from one tree to another, using their tails for balance. This is a primate that lives almost exclusively in these sixty square miles of Bhutan and northern Assam. Its population is stable today and has even increased dramatically in Assam because the villagers are intent on reforesting the land and being guardians of India's rarest primate. They know that when the langurs thrive, they thrive.

Hishey decided we needed some "forest bathing," a hike through majestic trees and over tiny streams blanketed with ferns. The "bathing" part comes from the increased oxygen the greenery releases and which the Bhutanese believe is instantly beneficial to well-being. It certainly felt good to be out in the lovely sun-dappled forest, and there were lots of songbirds greeting us, as well as the mournful piping of a Hill Partridge hidden in a grassy clearing. Suddenly we heard a rush of air from large beating wings as two huge Rufous-necked Hornbills landed noisily in a tree above. Then they glided to another tree and began tearing into some yellow fruit with their heavy beaks, a third again as long as their heads. Two stunning males sported the rufous-colored neck, head, and belly, contrasting with glossy black wings and a tail tipped with white. The color around their eyes was a startling turquoise blue, and their long white beaks were marked with vertical black stripes, suggesting applied war paint. It is said that you can tell the age of the bird by the number of stripes. I counted seven on one of them. Soon they were joined by four more hornbills. This was a great sight; the birds are threatened throughout their range and are extinct in Nepal and very rare in Thailand, where I had seen one before. But here in Bhutan they had not been hunted for their feathers nor their beaks. Here they were threatened by the logging of the big old deciduous trees they needed for nesting and food. Within a few days we saw seventeen of these remarkable birds, some flying over mountain passes at twelve thousand feet. The oddest sight of all was of a male Rufous-necked Hornbill sitting on a branch clutching an uncommon Parti-colored Squirrel in its beak and beating it to death against the trunk of the tree, its neck thrashing from side to side repeatedly. So much for thinking the hornbill is a fruitarian!

It was deeply satisfying to be in a place on earth where mammals, birds, and insects are held sacred by human beings, or at the very least respected as sentient beings equal to us. I wonder when we became killers. I can understand killing for sustenance and killing out of fear of being killed. There are spearheads in the bodies of Mastodons that roamed the Seattle area more than 13,500 years ago.

Some scientists even believe we helped drive these early Elephants to extinction. When did we begin to kill for sport and commerce? Did it arise out of warring with other peoples? The need for trading animal parts? Is our nature basically bellicose?

In 1959 when I was a young student just arrived in Britain to study for my junior year at the University of Edinburgh, I made contact with an English girl I had met the previous summer in the United States. Little did I know she lived in a castle in the Midlands dating from 1066, and that I would be invited to go on a "shoot" that Saturday with her knighted father and a few dozen others. The gentlemen, outfitted in tweed and herringbone, spread out in a line on the ancient fields, their guns at the ready, while "beaters" pounded the bushes to flush the birds. Pheasants, quail, snipes, and songbirds rose with a stunned swirl into the sky and shotguns blasted away. At day's end two hundred dead birds were packed side by side down the length of the driveway, testament to the virile exploits of the afternoon.

I was horrified but said nothing. I was nineteen and adhered to the maxim "When in Rome, do as the Romans do." A kindly, eccentric old lady, with a silk purse trailing off one wrist, approached me in the great hall when everyone was sipping sherry and downing congratulatory pints. She smiled sweetly. How was I enjoying my time so far? I ventured that the birds seemed to have no chance and that the number of them killed seemed excessive—far more than we could possibly eat. She whooped a bit, in sympathy with my tender perspective: "Oh, blood sports, my dear, that's all that matters."

I never forgot it. She seemed to encapsulate it all in that single statement: this is how men are, this is what excites them, get used to it.

I am not a softie. I do not wince when animals or fish are gutted; I have caught and gutted a fair share myself. And I am not a vegetarian, while I applaud those who are. Yes, I am conflicted. If all life is sacred I would not be eating or killing anything with eyes. I would not slap at blackflies or mosquitoes. So I am as predatory as the next man or woman. But I would give the animals a fighting chance—a sporting chance. In this day and age when so many species are in decline and fully 50 percent of our great mammals are threatened with extinction the balance is uneasy. My own predatory nature has been tempered through the decades by an understanding that we are all part of the family of things. Or as John Muir put it, "When we try to pick out anything by itself, we find it hitched to everything else in the universe."

The Bhutanese people seem to exist in harmony with nature. Is this because of a thousand years of Buddhist teaching? Is it because the king does not allow hunting or fishing? The king has also decreed that native dress must be worn in public places. The men look handsome in their short skirts, or ghos, and the women very beautiful in their woven kiris; it certainly adds to the allure of the country for tourists. But the younger generation is chafing at these restrictions, particularly under the new democracy and in the main city of Thimpu.

I want to believe the Bhutanese are kinder and gentler than the rest of us. It gives me hope that we can save the threatened creatures of the planet, that we are evolving as human beings and can coexist with wildlife in the future. But with a burgeoning global population I have grave doubts.

Even in Bhutan the specter of massive development of hydroelectric power threatens the most vulnerable of species. One bird on the top ten world list of those most threatened with extinction is the White-bellied Heron. We drove along the Puna Tsang Chu River up to the area of Tsekha village, where some of the last two hundred of these herons live. They prefer sandy gravel beds in the middle of the river, and as I peered through my binoculars I saw not one, but three of these tall, graceful birds resting on an island. They look like our Great Blue Heron but with a white belly and slimmer neck, and they have a longer beak. While the Great Blue is ubiquitous in North America, adapted to salt- and freshwater wetlands from the far north to the tropics, the White-bellied Heron exists only in these specific habitats on Himalayan foothill rivers, and nests in big chir pines along the banks. It is declining rapidly due to degradation of these sites, such as gravel removal from the rivers, and increased human traffic. And damming the rivers for hydropower causes the water level to rise and makes it harder for the Herons to catch fish. As they search farther afield for food they leave their eggs and young exposed to predators like eagles.

The White-bellied Heron is one of twenty-five global species selected by the organization Save Our Species for immediate attention. We passed signs posted on nearby roads urging the government to "Save the White-bellied Heron." The placement of dams and educating people about the unique status of this bird might make the difference. I invited some children to take a look at the Herons through the telescope. A little girl peeked through, and when she finally caught sight of the birds, she pulled back with a smile. I do not believe she had ever seen them before, even though she lived in a house just thirty feet away. If everyone had the gift of binoculars and scopes, the world would expand instantly.

Bhutan's leaders are trying to balance the needs of development with those of a sustainable environment. Prime Minister Jigme Thinley was on his way to see the site of a new hydro station and stopped for some meetings at the hotel we stayed at in Trongsa. My friend Hope Cooke had plied me with gifts to give to the PM and members of the royal family.

In 1963 the American Hope Cooke became queen of Sikkim, east of Bhutan, when she married the king, or Chogyal, Palden Thondup Namgyal. Hope and I were fellow students at Sarah Lawrence College in the late 1950s and later shared an apartment in New York City with two other collegemates when I was beginning my career as an actress. In the summer of 1961 Hope went to Darjeeling, India, where she met the future king of Sikkim in the bar of the Windamere Hotel. So began a magical saga of romance, royalty, and revolution.

The wedding was a much-celebrated affair, augured by monks who had determined the most auspicious date. It fell at a time when I was performing in a play and could not leave for the weeklong ceremony halfway round the world. Photos were splashed over the pages of *Life* magazine and Hope's beauty in her Sikkimese costume was heralded far and wide.

During the course of her years as queen, Hope and her king visited his royal cousins in Bhutan, and Hope became particularly fond of the queen mother of Bhutan and her sister. Sikkim and Bhutan were the last two monarchies in the Himalayas—related families that had ruled for hundreds of years. Nepal fell to the Maoists. India threatened to take over Sikkim. Ultimately, in 1972, Hope fled with her children to the United States, never to return.

When Hope heard I was going to Bhutan she followed custom by having me take gifts to the former queen and the prime minister. But she did not tell me how I was to find them. "Everyone knows," is all she said. Hishey contacted his cousin, who was a local justice near Trongsa, and he talked with the PM's chief of staff. Soon I was given an audience of ten minutes with the prime minister. He was pleased to hear about Hope and also about all the wildlife we were seeing. He said he loved nature, and hoped to leave an environmental legacy when he left office. I mentioned how special were the White-bellied Heron, the Black-necked Crane, the Rufous-necked Hornbill, and

With the prime minister of Bhutan, a man trying hard
to balance development with environmentalism

the Ibisbill, which also relied on gravelly riverbeds. How lucky Bhutan was to have these threatened species! Yes, he nodded. He didn't say that there would be sacrifices made, not when the future relied on the damming of rivers for hydropower, but it was what we were both thinking. It is not possible to save everything. Progress, you know.

I thanked the prime minister for his time and gave him the gift from former Queen Hope. He gifted me with a handsome Coronation Coin of the young King Jigme Khesar. I rejoined my fellow travelers in the great six-hundred-year-old dzong, where the monks were celebrating the cleansing of spirits from the temple by dancing all day in elaborate costumes and scary masks. They were stunning in their bright blues and reds and golds as they whirled and whirled all day in the courtyard to drumbeats, bells, and horns. This dance has been danced for centuries. It is possible to save dances and costumes and grand buildings like the dzong. Is it not possible to save the White-bellied Heron? The words of the naturalist William Beebe played again in my mind like a mantra: "The beauty and genius of a work of art may be reconceived . . . but when the last individual of a race of living things breathes no more, another heaven and another earth must pass before such a one can be again."

Hishey had a treat for us. One evening while we were eating our usual fare of rice, eggplant, cauliflower, and an addictive chili-pepper sauce, he said that whoever wished to rise before dawn and drive up a mountain to a small monastery at eleven thousand feet might be rewarded with the sight of a Monal Pheasant. He knew of an old monk who fed the birds as they came out of hiding at dawn and dusk. I could hardly contain my excitement.

We wound through frost-covered fields, the backs of horses steaming in first light, roosters crowing the new day. Then we ascended for half an hour, arriving as the sun exploded over the mountaintop, spilling onto the red roofs of the monastery. We had barely alighted from the van when I saw the robed figure of an old man scattering seeds before him in a small barnyard. And there at his feet were eight Monal Pheasants.

My heart was racing. I moved in slow motion with my Nikon at the ready. The group split in three directions, some to a wooden balcony, some to the edge of the monastery, while four of us crept up a little path to look down on the scene. Then the most extraordinary thing happened. Two male Monals moved across the tiny barnyard followed by a few hens. They hopped up onto the grassy path from below and made their way, pecking at the soil as they went, toward us! We did not move a muscle. The first cock pheasant went from the dark cloud of the building into a shaft of sunlight, and every feather radiated iridescence. It was electric. He simply glowed. The aqua rim of down surrounding his black eye morphed into pale metallic green, then poured into a deep turquoise to produce teal as it slid down his neck, hitting a bright ring of copper ruff and then a deep purply back. An improbable blue topknot spurted straight up out of his skull in the shape of tiny spoons. It is said that the Monal Pheasant sports thirteen different colors, which need the play of light to be truly seen. The bird sparkles when touched by sunlight.

I stood motionless except for the click, click, click of my Nikon. I lowered the camera from my eye and watched amazed as the bird came closer and closer and finally moved right into my shadow. It was as if I had summoned him. Which I guess I had; he was the one I most wished to see. He pecked near my shoes for a minute or so and then turned, gathering his flock behind him and disappearing into the dry grass up the hill.

Hishey was below in the courtyard. I came down and embraced him in gratitude, shaking with emotion. We were all jubilant. The day was crisp and bright, the monastery was peaceful and timeless, and the glory of the Monal Pheasant had been perfectly revealed.

Our time in Shangri-la was coming to an end. Our last days propelled us into the urban mainstream of city life. Thimpu, the capital of Bhutan, is suffering from pollution, unemployment, teenage restlessness, and a drug trade. It is also the seat of a nascent cultural scene that includes filmmakers, writers, and visual artists. Archery is the national sport of Bhutan, and we watched a contest between two top teams. The archers danced like cranes each time an arrow pierced the target.

Thimpu is also where the "Ashis" live, the grand queen mother and her ninety-year-old sister, both of whom are known as Ashi. Serendipitously I had met Dasho "Benji" Dorji the day before. Benji founded Bhutan's Royal Society for Protection of Nature and is a cousin of the queen mother. He arranged for Ed and me to present Hope's gifts.

The palace looked more like the modest estate of a Greenwich, Connecticut, investment banker than the residence of Himalayan royalty. Made of stone, it is set on a rise and landscaped with flowers and shrubs you might find in New England.

The queen mother, grandmother of the current King Khesar, is a delightful, thoughtful woman in her eighties. She was dressed in a beautiful kiri of maroon silk and a jacket of turquoise and gold. She presided over a formal English tea, replete with crumpets and delicious lemon tarts. We talked of the wildlife we had seen, including the odd-looking Takin at the zoo that afternoon. The Takin, on its way to extinction in the wild, is a dark ungainly ungulate that looks like a cross between a cow, a black ram, and a moose. It is the national animal of Bhutan and proliferates happily behind zoo bars, but not elsewhere.

A Takin. Almost at an end in the wild, because of overhunting, it thrives in the zoo in Bhutan.

We presented Queen Hope's gifts to the Ashis and they gave us a beautiful book on the flora of Bhutan in return. Then the power went out. The power goes out with regularity in Bhutan, but I did not expect it to in the palace. Servants immediately brought illumination in the form of very modern battery-powered lanterns and we continued our tea as if nothing had happened. It is one of the charms of Bhutan that everyone rolls with the punches.

Then the conversation turned to a love the queen mother and I shared, that of Hope Cooke's daughter, Princess Hope Leezum. "Little" Hope, as we called her, and her older brother, Prince Palden, first visited our Putnam County home as youngsters, swimming and playing games with our sons Tony, Geoff, Jace, and Jon, who were close to them in age. Palden was a storybook-handsome prince and Princess Hope was a darling, feisty little girl.

In the early 1970s India initiated a takeover of Sikkim, as China had of Tibet. A strategic pass to China in the north was important for India to control. The monarchy began to collapse and Queen Hope, fearing for her children's lives, fled to the United States. A few years later the king died of cancer, India incorporated Sikkim as its newest territory, and 333 years of the Namgyal dynasty came to an end.

"Little" Hope graduated from Georgetown University and was starting a career in New York City when she and my stepson Geoffrey Sherin, who was attending the Culinary Institute of America, became reacquainted in a Midtown bar. They had last seen each other as adolescents; now in their twenties, they fell head over heels in love. Their wedding, on the sprawling lawn of our home, was attended by many Buddhist monks chanting sonorous prayers for the young couple throughout the steamy August afternoon. The future seemed bright. But while they were on their honeymoon in Sikkim, which Princess Hope had not seen since childhood, she found herself irrevocably drawn to her native land and its people. She felt compelled to live there and to help the Sikkimese hold on to their culture and language, much as her mother had done before her. Geoff felt he could not commit to a life as a chef in Sikkim, or as prince consort, and these two beautiful star-crossed lovers parted ways.

And so the queen mother of Bhutan and I talked fondly of Princess Hope and all she was accomplishing in Sikkim to keep the ancient culture alive, despite the fact that she no longer had the infrastructure of monarchy that Bhutan had.

It is a time of great transition for this unique country. Poised to become a leading source of water and hydropower for the masses to the south, Bhutan will have to make hard decisions about its future. The great rivers of the Himalayas are the lifeblood of India: the Indus, the Brahmaputra, and the Ganges. As the world warms and lowland streams and tributaries dry up, as grasslands and paddies turn to desert, the Himalayan melt feeding the rivers will become increasingly important to more than six hundred million people. The mountain countries know this and are planning to build hundreds of dams in the next few years. Sikkim alone has twenty-two hydroelectric stations in the works, and Bhutan has triple that. They are banking on the economic prosperity that will ensue, but the downside may literally wipe all of it out. No one can predict how fast the glaciers will melt. Right now the Himalayas, with more than fifteen thousand of them, seem safe at the highest altitudes. But there are hundreds of glacial lakes, which are increasing in rate and size as the glaciers retreat annually. These lakes could burst their frozen walls in just a few minutes, triggered by water pressure or earthquakes; the entire area of the Himalayas is seismically active, as the massive earthquake of 2015 in Nepal attests. Fully 10 percent of the existing glacial lakes are currently in danger of having such an outburst flood. Any dams built will have to withstand unprecedented pressure in order to hold.

Bhutan also has "the worm." *Cordyceps sinensis* is not actually a worm but a fungus that grows on a caterpillar. The desiccated carcass with the finger-like fungi on it is allowed to be handpicked from below the ground in late spring. One kilogram of "worms" can fetch as much as $20,000 in the thriving Southeast Asian medicinal market. Cordyceps is touted as an aphrodisiac, a boost to the immune system, and an antidote to liver disease and certain cancers. Like the days of the gold rush the valleys where the worm lives are overrun with pickers turning over the soil. So far Bhutan has been able to manage this boom business. When the soil is replaced after disturbance the worms continue to breed. It is a well-balanced natural resource. And there is tourism, limited today so the country will not be overrun and because of insufficient infrastructure, but that will be developed. With exceptional leadership Bhutan will still manage to get many things right. It is the first nation in the world to ban monoculture and go completely organic with its agriculture. It is also carbon neutral; emissions from agriculture, energy, industry, and waste are offset by forests, fully 70 percent of the country.

The power came back on and we bid goodbye to the gentle Ashis. The queen mother stood at the door as resplendent in her silks spilling to the floor as the rare birds we saw in their luxurious feathers. The new young queen plays basketball in shorts and a T-shirt with a local girls' team. Things change. As democracy becomes more established, people will demand more development and the opening of all kinds of new global markets straining the peace of this peaceable kingdom.

Buddhism counsels against attachment, because nothing is permanent.

Robert Browning wrote, "If you get simple beauty and naught else, you get about the best thing God invents." Bhutan is beautiful. Its Black-necked Cranes are beautiful. My eyes misted at the thought of the glorious Monal Pheasant. When the old monk dies, who will be there to feed him in the morning light? How long can he escape the collector's net? And will such beauty come again?

21

Bahamas

Three floating skyscrapers with names like *Enchantment of the Seas,* *Norwegian Getaway,* and *Disney Dream* were attempting to come about in a harbor the size of a few football fields. They were each a thousand feet long and twelve stories high, and the displacement from their combined 400,000 tons threatened a tsunami on the lawn of my hotel. It was dawn; I had found a yoga mat in a closet of the ubiquitous exercise room where TV blasted to the grind of treadmills and clinking weights. My down dogs and warrior poses were thankfully done, and I ducked outside between tropical showers to watch the considerable skill it took the captains to negotiate the turns. It was a slow process, one ship following the next in a precise ballet they must have executed a hundred times. A small wake washed over twenty feet of lawn, and that was it. The ships, each with thousands of passengers tucked in their berths, powered out of the harbor and headed for Coco Cay.

Nassau on Providence Island in the Bahamas is one of the most popular cruise destinations in the world. The ships are playgrounds featuring gambling, dancing girls, spas, and pools, then traded on arrival for more gambling, more spas, steel drums, water sports, and high-end shops. If a cruise is not your thing, you can fly to Atlantis, a self-contained resort city with a thousand rooms offering casinos, spas, dancing girls and boys, water sports, and the added attraction of swimming with dolphins. And if resorts of this size don't interest you, there is some of the best fishing in the world to be had in the clearest waters you've ever seen. The Bahamas cater very successfully to the tourist trade, which employs half the population, making it the third-richest country in North America after the United States and Canada.

There are about two million islands on earth. Most of them come and go with tide and time; about 180,500 are estimated to have any appreciable landmass. Greenland is the biggest, followed by New Guinea, Borneo, and Madagascar. In the United States, Alaska has the most islands, with Florida and Michigan vying near the top—yes, Michigan, with all its lakes. New islands are formed from volcanic activity along oceanic continental plates, from coral reefs above the water making atolls, and from barrier reefs close to shore composed mostly of sand. And there are myriad islands in lakes and rivers that are part of the landmass.

I have gravitated to islands my entire life. There is something comforting about confines, the belief that you can know the nooks and crannies of an island intimately within its watery borders. Fifty years on the barrier island of Nantucket certainly seduced me into believing I knew the length and breadth of its thirteen by three miles. My parents bought a small cottage on Surfside Beach in the early 1950s. There were a handful of old fishing cottages on Surfside back then, and our days were spent barefoot and careless, swimming, sailing, clamming, crabbing, or casting for Bluefish when the run was on. I would head out to Eel Point to see Short-eared Owls cruising the dunes for rodents, or Glossy Ibis seeking baitfish in the pond. Least Terns and Piping Plovers nested on select beaches. I knew the best places where Blue Crabs hung out in the salt marshes, where the tiny Northern Saw-whet Owl roosted, and where a few huge white oak grew in a hidden copse. I kept my secrets to myself, but small islands really have no secrets. People have been stumbling upon them for eons, and changing the equation with each revelation. We can never know anything with complete assurance because the landscape is always changing, as axiomatic of our relationships with human beings as of the land. Nantucket's sandy shores were always moving; the ocean took 120 feet of our land in one storm back in 1974, leaving the cottage just 12 feet from the bluff. The ocean built the island five or six thousand years ago and the ocean will take it away with the rising tide of glacial melt in another five thousand years . . . or less.

Through the 1980s and '90s Nantucket became the darling of a new class of wealth, people who built mini-mansions with manicured lawns to the waterline, paved the dirt roads, and packed the harbor with yachts and the town with boutiques. The scallop beds diminished in the runoff of pesticides, clam beds were dug out, the birds were disturbed, and the line of jeeps fishing for Blues off Great Point was miles long. Surfside's few cottages, which looked east to the sea and west to fields of grass, heather, and scrub pines, were now crammed to the bluff's edge as hundreds of homes piled in behind us. The deer and the owls disappeared and the rare indigenous tulip-like lily was crushed underfoot or picked by folks who had no idea there were just a few dozen left in all the world.

This is an island story told many times in many places. Beauty is a magnet. Everyone treasures the experience of it, but there is a cost: the very elements that make a place beautiful, the wild land and sea and the wild creatures that live there, are sacrificed to the demands of human beings. And we are rapacious, as predatory in our needs as any species that has ever roamed the earth.

The Bahamas is a country of seven hundred islands, some underwater at high tide but more than thirty of them inhabited. It was here in 1492 that Columbus is reputed to have first landed in the Americas, although the Dominican Republic also claims the distinction. Not actually in the Caribbean, the islands occupy an area of 186,000 square miles partly in the Bermuda Triangle, 50 miles southeast of Florida. The climate is balmy year round. The waters surrounding the Bahamas are teeming with aquatic species. The islands sit atop three banks of fossilized corals and shells at the edge of the continental shelf, giving the shallow waters a stunning turquoise hue where people play, fish spawn, and birds feed. Fifty thousand West Indian Flamingos live at the edge of shallow lakes, while in the deep waters off the shelf great ocean fish ply the canyons for their next meal.

Migratory birds, abandoning the cold North, alight in the Bahamas for the winter just like human "snowbirds." Congregating by the thousands on remote cays and lagoons, shorebirds poke their beaks into sand and silt for tiny worms and snails.

The Piping Plover is one of those birds. The handsome pale plover with a partial black collar and a chunky little breast has never been a bird easily seen. Because it fades into the coloration of the sand, it goes unnoticed, and except for piping in distress, it is silent. It breeds from the coasts of North Carolina up to Prince Edward Island and Newfoundland, and also along the shores of the Great Lakes. The plover pair scratch an almost-imperceptible indentation in sand and pebbles above the high-tide line near a wrack of seaweed and plugs of dune grass, and both sit on the eggs until they are hatched twenty-five days later.

There are about 8,000 of these small plovers left, 3,100 on the Atlantic Coast, where I first encountered them in childhood. I remember picking up a chick and putting it in my sand pail on a Nantucket beach when I was about five. How could I not? It was the most adorable puffball of a toy I had ever seen, and I carried it about in ignorance of its mother piping her alarm call twenty feet away. My own mother was not so ignorant and gently chastised me, placing the tiny one back on the sand, where it scurried back to the nest scrape and under its mom's soft belly.

There must have been hundreds of children on hundreds of beaches throughout history who have done the same. The plovers share our coastal territories and suffer many indignities, onslaughts, and death from human and animal disturbance. Beachgoers leave trash, attracting seagulls and crows, which pick off the chicks. Jeeps run over them, and dogs scatter them. Too many people walk too close, eventually weakening the birds' resolve to lay eggs, or to raise their brood of three or four. There is predation by raccoons, foxes, owls, feral cats, and Coyotes. It is a wonder that any of the chicks fledge at all, but half of them manage to every year. The Piping Plover is a plucky bird.

I am a "Piping Plover Guardian" in my southwest area of Nova Scotia. Bird Studies Canada recruits volunteers to monitor the beaches for the birds and educate people about them. We observe when the birds arrive to scout nest sites, we check the eggs and chicks regularly to make sure they are safe, and we keep track of disturbance, predation, and abandonment of the nests. It is a joyful and sad task. The joy is in seeing these chicks pop out of the nest just hours after birth, looking like chorus girls in tutus with their long slim legs and puffy down skirt, scurrying about to forage for invertebrates. The sad part is seeing the tracks of a fox or a gull when the nest is empty.

My favorite beach to monitor is a wild windswept expanse of crashing surf locally called Hemeon's Head. I love the long walk through the grasses, despite the multitude of ticks, and the high berm of cobblestones all the way down the high-tide line to the flats of Matthew's Lake, a mile away. Fledge success has been difficult at Hemeon's because of so many predators. The year 2014 was cause for cheer.

In the seventeen years I had been a guardian on our southwestern beaches I had not encountered a pair of plovers I found more endearing.

The female was particularly beautiful, a very pale sandy-beige color, the black collar around her neck widening to perfect triangles beneath her head. When I first saw her on May 30 she displayed nesting behavior—agitation and diversionary tactics—before she settled down again on the same bare scrape of sand between several stones. Later in the day she did indeed lay five eggs, one more than is usual.

I saw the male on June 16. He was more traditional looking, a darker beige body and a black collar all the way round his neck. He too was calm, standing on the cobble like a sentry. I watched the changing of the guard as she came off the eggs and went to feed at low tide and he took over incubating.

On June 28, right on time, I saw the dad on the nest with a fluff-ball at his breast, the first chick. Within twenty-four hours two other chicks had hatched; the remaining two did not.

Hurricane Arthur blew in on July 5 with 70 mph winds for four-teen hours straight. The bird fallout when it all subsided at 7 p.m. was huge. Laughing Gulls fell out of the sky by the hundreds all up and down the coast, and Black Skimmers were found on Cape Sable and the Eastern Shore. Following the previous hurricane it had rained blue-colored birds: Indigo Buntings and Blue Grosbeaks!

I felt sure the plovers were in trouble, and first thing the next morning I raced down to Hemeon's through washed-up kelp and storm flotsam to their nesting area. I heard and saw nothing in the wrack and pebbles near the nest. I sat on the cobble to deliberate. They weren't on the beach feeding, nor near the nest site. Then I heard the briefest of piping, and the pale female flew low out of the wrack and landed on a pile of stones one hundred feet away. Then the male, standing guard beneath the "Piping Plover Nest Area" sign, followed and sat next to her. For a moment I thought they were going to copulate, which made me believe the chicks were dead, but no, they acted as if they were signaling the chicks to stay put. I rose and nonchalantly walked down toward the water a bit south of them.

That put them at their ease, because Dad took off flying to the far eastern end

A Piping Plover chick tugs at a worm, hungry after the hurricane in 2014 in Nova Scotia.

of the beach to feed, while Mom stayed put on the stones. Then the tiniest of heads peeked over the cobble next to her and a sweet fluffy little body emerged and made a beeline for the water's edge. Then a second one a bit smaller took off like a windup toy after the first, which caused Mom to take after them with cautionary piping. I was so relieved they were all right, snapping pictures with my long lens, that I didn't at first see the even tinier third chick appear on the cobble as if no one had woken her up. She spun her little stilt legs toward the shore and joined her siblings slurping up slim black marine worms like spaghetti while Mom stood her ground, a terrier between us.

The female left by July 16, off to mate and try to nest again, leaving Dad to bring up the chicks. When you've got only a few thousand of your species left you grab every opportunity you can to procreate. I think these chicks survived because she and the father were so attentive together for so long. They never resorted to distractive behavior to divert attention from the chicks, but rather hunkered the chicks down and observed my intrusion until it was safe. Very smart. Predators such as gulls, owls, and crows were not alerted to the chicks' presence.

I walked up over the cobble and gazed at the grasses as far as the eye could see, extending to the lake and to Ram Island offshore where a little colony of Puffins breed. It was a perfect day, and here I was all alone with the birds. Then down the path came Russel, a birding friend, with his big 400mm lens.

We birders are so predictable! We think alike and can be found congregating in the most remote places at odd times. It is comforting. I suppose every subculture has a similar agenda. We talk shop: What have you seen? Where? When? What's coming up? Want to go on the pelagic trip in August?

As Russel and I chatted, a pair of Gull-billed Terns circled the small pond, fallouts from the storm, their pure white bodies contrasting with a cap of jet black and a big black beak. I told Russel where the plover chicks were if he wanted to get some good shots with his big lens, but he demurred, saying he didn't want to disturb them when they were so young. Now *that's* a true bird lover.

The chicks fledged and took off for the South some time after August 3. All Piping Plovers, abbreviated as PIPLs, were banded in Nova Scotia in 2014 by Fish and Wildlife. Nova Scotia was doing a great job keeping the plover count high but fewer were returning the following spring from their wintering grounds, and we needed to know why. The dad of my pair had a gray flag band with the letters "HE" on it. He took off in early August and was reported at North Brigantine Natural Area in New Jersey on August 6. He's a winner. We call him HEmeon.

There are guardians or "stewards" of these birds almost everywhere plovers are found these days. On the West Coast they monitor Snowy Plovers, which are also in danger. It makes a difference. The public has responded to signs educating about the plovers and asking that they walk at the low-tide line with their dogs leashed. Most people do not mind that portions of the beach are closed during nesting season, although there is an ongoing ugly war in North Carolina and other southern states, with jeep owners wanting beach access at all times.

We know a lot about these birds now, and just when we think they might not make it, we are surprised by their tenacity and the enduring generosity of the public, who want to see them make it. A popular spot in southwest Nova Scotia, Louis Head Beach, had two nesting pairs a few years ago. The beach was frequented by folks in rental cottages and in a nearby trailer park. There were many dogs frolicking in the water and over the beach grass. We didn't think the plovers had a chance.

The guardians of that beach took it upon themselves to talk to everyone, pointing out the nests and how fragile the situation was. The dog walkers were asked to take special responsibility—a rope marking the nesting area was no deterrent to a roaming dog. One nest was right at the end of a short boardwalk from cottage to sand. The guardians showed the vacationers the nest and the four eggs and explained the timeline of incubation to fledging, about two months. They lent the family a pair of binoculars and encouraged them to be on the lookout for crows, seagulls, and mammals seeking a meal. The little children were awed by the nesting birds they had somehow missed when building sand castles, and vowed to protect them. I watched from a distance one afternoon as the seven-year-old boy made a wide arc around them on his way to the cottage, never using the boardwalk at all. His pride in aiding these small lovely creatures said it all. The family became the best guardians we had ever had, proud of the four fledglings they protected. They became dedicated bird lovers from that time on.

There is a story of two fellows walking a New Jersey beach strewn with plate-sized Horseshoe Crabs that had washed in on a high tide and been flipped on their backs, leaving them immobile. One of the fellows begins to turn them over, giving the crabs the chance to inch toward the water. "What are you doing, Mike? There are hundreds of them, what difference does it make?" And Mike replies, "It makes a difference to this one." That is how things change: one by one by one.

Piping Plovers spend only a few months on their breeding grounds before they start their journey back to their winter homes, stopping at known fueling spots along the way, just as we might stop for a meal. If there is degradation of the sites or depletion of food sources the birds will not be able to double their weight as they need to for trips over water. Every stop counts, and the juveniles are most vulnerable, learning as they go.

Fully 10 percent of the Piping Plovers make it to two remote cays in the Bahamas. This was only discovered in 2011 by scientists doing a survey; they counted 350 of the endangered birds hunkered down on the white sand among thousands of other shorebirds.

In March 2014, a few of us on the board and staff of the National Audubon Society and the Bahamas National Trust hopped on a bonefisherman's boat and arrived an hour later on one of the cays to witness the birds. There they were, sweet Piping Plovers bunched together under the warm trade winds like sunbathers on Coney Island. For those of us who had never seen more than a few birds at a time, this was a deeply rewarding sight. Some of them were likely the very birds we had monitored up north.

In the United States we call them our plovers, but the Bahamians rightly have more claim, since the birds spend as much as eight months with them. In any case it takes all of us along the flyway to usher them safely back and forth. And this is what we on the Audubon board were there to ensure. David Yarnold, CEO of Audubon, and Kenred Dorsett, Bahamian minister of the environment, signed a memorandum of understanding to protect the shorebirds wintering in the Bahamas, in hopes that these particular cays would become part of the park system of the country, safeguarded from development.

The Bahamas are particularly vulnerable to development, since there are so many beautiful islands and so few people. They have managed to keep big popular resorts like Atlantis close to the city of Nassau on the island of New Providence, and most tourists disembark there from planes or one of the huge cruise ships that arrive daily in port.

The Chinese discovered the islands and started to build an extensive resort near Nassau with enough hotel rooms to house as many as twenty thousand future Chinese tourists until they ran into financial difficulties in 2015. The Bahamas National Trust, the environmental arm of the government, preempted the Chinese desire for shark fin soup by banning all shark fishing in the Bahamas in perpetuity. Sharks are to the Bahamas as House Sparrows are to New York City, so this was a wise move. We all know what happened to the most plentiful bird in North America a hundred years ago: the Passenger Pigeon was hunted recklessly to provide the upper class with "squab" for dinner.

Most tourists are happy to stay on the lovely beaches of New Providence, while away the hours in the casinos, see the aquatic shows, and frolic in the pools. They do not feel the tug that fishermen, divers, and birders do to explore the natural resources of the islands. And this intrinsic balance in human interests is what will keep places like the Bahamas from being degraded, as long as the government regulates where and what kind of development takes place, and as long as the youth of the country are taught to cherish the wildlife of their remarkable islands. For now, the hardy Piping Plovers, snowbirds escaping the cold North just as we do, are safe on Bahamian sands.

In 2015, HEmeon, the plover dad I monitored, returned to the same area on Hemeon's Head, mated with a new female, and sat on four eggs until predators took them. The mom of 2014 found another mate on a beach farther south. We still do not know the fate of the three chicks they successfully fledged from the summer of 2014, but I like to believe they made their inaugural voyage over the ocean to the white sands of the Bahamas and back again to some secluded beach somewhere in the Maritimes, where they are raising their own chicks.

22

Ecuador

Wavy blobs of color permeated my feverish sleep. It was hard to tell what they were or if there was any substantive story to be told, and I was helpless to form a story if there was one. By the time the taxi dropped us at the little Quito hotel, I knew I was ill. Although I would have liked to ascribe my fatigue to one of the tropical fevers I invariably got when I traveled near the equator, I knew I had brought this one from New York. I collapsed in my narrow bed and woke only once in the next twenty-four hours.

My traveling companion, Julie, had hardly ever been out of Nova Scotia in her life. She was, fortunately, resourceful and ventured out to the plaza on her own to see the people in the fading evening light, the small sturdy women in their porkpie hats and musicians with their ethereal panpipes. But within the hour the altitude got her. At nine thousand feet Quito can be a shock to the system of someone who has always lived at sea level.

By dawn the next day my fever had broken and cup after cup of mate tea had quelled Julie's headache and queasy stomach. We boarded a small van with two Norwegian medical students and began our trip south through the volcanoes to the little town of Shell, founded by the oil company in the 1930s before exploration was abandoned. Then an hour flight in a single-engine Cessna across unbroken forest terrain over veins and oxbows of the great Pastaza

River, making its way to join the Amazon. The airstrip was a short mudpack of a runway. A few children with their mothers rested under a thatch of shade and watched as we slipped into motorized canoes for the thirty-minute ride to Kapawi. We arrived at the clan's eco-lodge in time for lunch, a trip that would have taken ten days had we walked from Shell. There are no roads to Kapawi except those by water.

The Achuar were making enormous changes in their lives. As one of the fiercest of the Amazon Basin tribes they had never allowed the Spanish near, earning a reputation as "savages" who speared, decapitated, and shrunk the heads of their enemies. They fought other clans as well, their linguistic relatives the Shuar to the west, and those to the north, each side taking enemy heads. The Shuar were greater in numbers and pushed the Achuar out of the higher elevations and into the fluvial lowlands surrounding the Pastaza. When the oil companies arrived to explore four hundred years later, their welcome was much the same. Shell retreated in 1948, considering the Achuar too dangerous, and things were quiet for the clans straddling the Peruvian border until 1964, when Big Oil returned with a vengeance. This time it had the law on its side.

Julie and I were given a spacious palm-thatched hut on stilts, screened against tiny intruders but open to the shallow wetland underneath that was called a lake because it rose with the rains. It was brisk with birds, the perky Blue-gray Tanager, the whooping call of Oropendolas, yellow and black Caciques, and a pair of Horned Screamers courting underneath the boardwalk, an endangered species according to Ecuador birding guru Robert Ridgley. Although there were enough huts to accommodate forty or so we had the place to ourselves except for a family from Oregon. It was quiet, only the natural sounds of birds, frogs, katydids, and a troop of Red Howler Monkeys filled the air. The howlers have a disconcerting roar that sounds like an approaching freight train. There was no vehicular traffic, no human voices, no machines. It was one of the remotest places I have ever been.

There was a bird I had always wanted to see and missed on prior

A Hoatzin, a puzzlement in science

trips to the Amazon. The Hoatzin is a strange one indeed: the jury is still out on how to categorize it. The size of a pheasant, the bird congregates in trees over water, looking like a throwback to earliest species of the Miocene or even the Eocene eras, which it may well be. It eats a diet of leaves, and Hoatzin chicks have claws on their wings so they can climb up branches. No creature eats the Hoatzin, as its flesh tastes terrible from all that leaf fermentation. Natives call it the stinky bird . . . not a bad survival skill.

We were cruising down the river in our open canoe that first afternoon when many of these fascinating birds squawked and grunted above us in the overhang. Their facial coloration was beautiful: bright red eyes, encircled by electric-blue skin down to the beak and a spiky chestnut mohawk on top. On a tree across the river a large Three-Toed Sloth wobbled in the highest branches seeking a place to rest for the next twenty-four or forty-eight hours. They don't get around much. A pink Amazon River Dolphin surfaced and then another, tumbling over each other in courtship. Julie spotted a

Lesser Anteater, with its prehensile tail, on the river's edge taking a drink. And a pair of sweet little Saddle-backed Tamarin peeked out at us from a log, while Squirrel Monkeys, Black-faced Monkeys, and Night Monkeys looked on from above. We threw lines in the water, but our catch of piranha was too small to keep.

After my marathon fever sleep in Quito I was awake most of the night, happy to listen to the creatures and look though the screen at a million stars. There was Orion's belt and the Seven Pleiades as in our Nova Scotia sky, but the Southern Cross, never seen up north, had not yet made her rounds of the Amazon Basin.

We had come to this place because Julie wanted to experience the mind-altering drug called ayahuasca. She read extensively about the ceremony and knew there were indigenous people for whom it was sacred. My experience with mind-altering drugs was limited to LSD back in the 1960s, and the trip I had taken had not been good. I had no wish to repeat that night. Hallucinogens, marijuana, opiates, and amphetamines were present throughout my college years and twenties, as they were for many of my era. Mostly I just took amphetamines, or "speed," as we called it, to get through a paper-writing night.

One of my best friends at Harvard volunteered to be a guinea pig for Timothy Leary and Richard Alpert in some of the first LSD experiments in the early 1960s and would report to me on the fascinating visions he had. One winter day I found myself skiing with Dr. Leary and some friends in Aspen, Colorado. Leary claimed never to have skied before. He certainly was not dressed for it—a weird pair of boots, an old pair of jeans, and a cloth jacket were not suitable at eleven thousand feet on Ajax Mountain, but he was doped up, and as we all stood together before takeoff, Leary looked around at the snow-clad peaks and the sun bouncing off them and proclaimed, "I have never been so high." Then he just went lickety-split down the long run without stopping or even falling until he hit bottom. "Turn on, tune in, and drop out" took on new meaning. I was impressed with his stamina, if not the nature of his mind, which got him down the mountain in one piece.

Drugs were not a big part of my life. I did not like the hangover, nor did I crave them. And I found I could not inhale and act or drop acid and go onstage as some of my colleagues were able to do. As I turned more and more to birding and the great outdoors, I found my high in wilderness and wild things. I was more than content to wander the jungle and canoe down the rivers accompanying Julie on her vision quest.

My friend Julie Balish from Nova Scotia goes native in the rainforest of Ecuador, 2014

The Achuar were totally connected to the forest. It was alive for them and everything had its name and sacred essence—the trees, the water, the vines, the rocks, and all the creatures. Some were more powerful than others: the Jaguar, the eagle, and the Anaconda. Like all the clans in this part of the world and two thousand miles north, where the Lacandon lived, and on into Native North America, dreams and visions were as important as waking life.

Dreams were interpreted in the predawn hours as Achuar families rose to begin their day, their dreams telling them what might happen and what to avoid. The Indians of the Amazon Basin have a rich spirit world. The Achuar each have their own *anents,* or magical songs, which they sing to spirits they wish to reach. A woman might sing to the garden spirit for a good crop, or her husband to the spirit of peccary to ensure a successful hunt. There are also spirit protectors that will avenge any disrespect of animals or the forest or when a person is greedy or mistreats animals. When a Jaguar calls, it is believed that someone in the family will die. The clan believes in many superstitions and myths, and nothing in the natural world occurs without note. This animism, uniting the spiritual and temporal worlds, used to be found in most Indian cultures of the Americas. It is deeply part of the Achuar still. Missionaries even wrapped

Christian stories into the existing cosmology of the Achuar, although the balance of animism and Christianity is uneasy. Arutam is the sacred source, the life force of the forest, rocks, the sky, the plants, and the animals. It is everything, and there are sacred places where Arutam resides. One must seek Arutam through a vision, and the guidance and wisdom one receives is then with a person for life.

A boy lets his father know when he is ready to receive Arutam. At ten or eleven, accompanied by his father, he selects a sacred place in the forest such as a huge kapok tree, where he clears an area for the Arutam to enter and erects a palm shelter for himself. He touches the tree, asking for ancestral guidance. When the sun is low he drinks a plant concoction, usually a tobacco drink, with the power to release visions, and he lies down to sleep, his father watching over him. It is then that the visions, hopefully, appear to the boy and he receives Arutam.

The visions can take the form of animals, frightening or comforting, or of water or mountains—all have their meanings. They represent ancestors, the boy's grandfathers or great-grandfathers, giving him spiritual guidance. These visions, this Arutam, is for the boy alone and must not be shared with others. He may not receive it the first time, but he will try again until he does. A young woman may go on a quest for her Arutam, but she can also receive it through her husband's experience.

If I knew nothing about the spiritual life of our Achuar guides at Kapawi, it would not be difficult to intuit their extraordinary connection to nature. We spent our days taking forest walks with Simon Santi, who painted bold black geometric patterns on his face daily out of respect for the forest. He was a quiet man and gentle, in his late twenties, his smooth skin a creamy umber and his dark brown eyes averted lest they seem aggressive or flirtatious. The Achuar avoid displays of emotion. He wore a headband of leather, mounted with small feathers of toucan and macaw in red and yellow. He spoke no English. Fernando, a guide from Quito, accompanied us to translate.

Simon might have had extra eyes, he was so skilled at spotting the tiniest tree frog splayed on a green leaf in perfect green consonance, or a mottled brown one in the forest litter at our feet. He cautioned us to be careful not to step on Leafcutter Ants bivouacking in line with their green flags to their huge underground home. It had several entrances, and a group of ants designated the quality-control squad intercepted the workers, taking away leaf bits that didn't make the grade, before they were allowed to enter.

Simon told us to beware the biting Army Ants and he let an imposter, the fake Army Ant, crawl all over his arm. I couldn't tell the difference. He showed us the "Dragon's Blood" tree with its antiseptic red sap turning to white when rubbed into the skin. Minuscule Lemon Ants patrolled the bark of a sapling; in symbiosis they kept other plants from growing nearby with their own herbicide in exchange for their tree home. They smelled lemony, and with Simon's encouragement I popped a pinch of them in my mouth. They went down easily and made my breath sweet and clean for the next hour. A red Poison Dart Frog smaller than my little finger sat on a mushroom, glowing like the noxious little creature it was. There was enough poison on its tiny back to paint ten arrowheads and kill as many animals, although the curare plant was used more frequently. The Achuar still hunt with long blowguns and unerring accuracy.

This forest was not very "birdy" for reasons I didn't fully understand. There *were* birds, but not as many as might be expected in Ecuador, which has more birds than anywhere in the world given its size, more than sixteen hundred species. Perhaps they had been killed, although the eco-lodge was in an Achuar reserve off-limits to hunting. If they had been killed before the reserve was established it could take years, as many as thirty generations, for a species to reestablish itself. Still, it was a puzzle. Except for the headbands, which are passed from generation to generation, I did not see the clan wearing abundant feathers, and small birds wouldn't make more than a one-bite meal.

Yet, at a salt lick on a river cliff, we did see multitudes of Orange-cheeked Parrots, Dusky-headed Parakeets, and Chestnut-fronted Macaws gathered for their breakfast vitamins, and on a sandbar a small flock of glorious Blue and Yellow Macaws took off and flew above our heads, while more than fifty Swallow-tailed Kites, refugees from the wintry North, wheeled over the river.

It was on night walks that the forest truly came alive. Our headlamps revealed many different kinds of tree frogs everywhere, the Great Tree Frog, as big as a Bullfrog, sitting on branches with its white belly lit up, and more Poison Dart Frogs, tiny little spots of blue, green, or yellow, waiting for an insect meal. Insects abounded, clouds of soft-bodied Whiteflies shimmered over pools of water while a pair of six-inch Walking Sticks in the act of mating blended into rough brown bark. A fist-sized arachnid they called a Scorpion Spider with an elongated body clung to a cistern wall. Luminescent beetles and more species of bats than I could count fluttered past us in the dark.

I loved the bats especially. There has been a huge decline of bats in the United States; a disease called white-nose syndrome has killed most of the Little Brown Bat population. This was the same species that I surprised on opening our attic door when I was a little girl and which became hopelessly tangled in my long hair. I was screaming, my mother was cutting the animal out, and my father was assuring me that everything was okay. He put the little bat in a box and we kept it for several weeks, Dad teaching us how harmless it was and how important bats were to the world.

When my husband and I moved to Putnam County, New York, we put up a bat house on the siding overlooking the stream. Before long it was filled with bats. There was no problem with mosquitoes in the summertime. There was no problem with insects inside our house, either. The bats found a way into the roof of our cathedral ceiling in the living room, some of them cozily clinging to rivets in the beams. One especially large bat I named Henry would occasionally swoop down to grab a bug out of the air while I was reading on the couch. I told guests he was harmless, to pay him no mind. Thirty-four years later we sold the house to a young couple with a baby who rightly asked us to get rid of the bats. We decided to put on a new roof. The roofer stopped counting at 120 bats when the old roof came off. There were plenty of big old trees in the woods with hollow trunks. I hoped they found a good one to live in.

Julie was preparing for the ayahuasca ceremony through fasting and repose when she heard a ruckus in the reeds beneath our hut. The male Horned Screamer was pinning the female down violently as he attempted copulation. Every time the female tried to rise, he beat her head and neck down with his wings until she was bloody and exhausted. This went on for forty-five minutes before she staggered away and the male flew to his usual perch on top of the highest tree, screaming territorially. I have seen male Mallard Ducks gang-rape a female in an equally violent act, but I did not know Screamers behaved this way. Neither did ornithologists I spoke with later.

In the late afternoon we hopped into our long canoe and journeyed an hour to the next village where the shaman was to guide the ayahuasca ceremony. A few children were splashing on the shoreline below the high red bluff where we put in. We climbed up the hill to an open and well-swept clearing the size of a football field and walked to the largest thatched dwelling in the middle. Ducking under the thatch, I saw a man sitting on a chair, holding a small mirror and applying black face paint with a brush. He was dressed in a pale blue short-sleeved shirt with the kind of headdress Simon also wore, a crown of small red and yellow feathers. On his chest was a necklace of finely patterned tiny beads that reminded me of the work of Sioux Indians. I thought I was backstage in the theater again. He looked as regal as Laurence Olivier, with his silver hair and handsome features. The ritual was the same, the concentration intense, and the transformation almost complete. He was soon to perform.

We sat down on a far bench and waited. The shaman, Sumpah, put down his brush and turned to us. He gestured to his wives in the far corner, who were huddled over a big pot. The eldest brought us gourd bowls of fermented manioc, or nijiamanch, the ubiquitous drink found all over Central and South America which goes by the common name chicha. This can be as strong as a good European beer or as weak as Bud Light depending on how much water is added. The Achuar drink it morning, noon, and night.

Sumpah asked each one of us, through Fernando, who we were, where we came from, what our jobs were, and why we were there. He radiated easy power, effortless intensity, as he listened to each of us.

The Norwegian couple impressed him with their love of medicine and the wealth of healing plants in the Achuar forest. He looked very intently at Julie when she spoke of her desire for his guidance under ayahuasca. And he spoke lengthily of the need for all of us to come together to protect nature when he responded to me and my love of wildlife.

Fernando translated and then elaborated more on the huge fight the Achuar have with the "Company," as development was called in the Oriente region. Oil and mining were decimating forests and changing cultures entirely as roads broke through pristine lands. The government of Ecuador had granted sovereign rights to native peoples all across the Amazon Basin, but those rights did not include minerals beneath the ground.

Ecuador made a satanic deal when it hitched its future to oil production from corporations outside the country. Many global corporations like Texaco (now Chevron) and Canadian and Chinese corporations invested heavily but did not bank in the country, which put Ecuador in massive debt, from which it has not been able to recover. Petroecuador now drills most of the wells, but the price tag is still too high, both economically and environmentally.

Ecuador is one of the biodiversity hot spots of the world, with more species of plants and animals than almost anywhere else on the planet. These places should be sacred. The people who live there believe them to be sacred. In its rush to development the government trampled on the environment and the health of its indigenous people. Thousands of oil spills from leaking pipes carrying crude to the coast, or the waste pits that swell during rains and run off into fields and streams, have been documented for forty years. The incidence of cancer has risen dramatically in villages near the wells, and downstream. The people call the disaster "the Chernobyl of the Amazon."

Missionaries in the late 1950s and '60s encouraged the hunter-gathering Huaoranis to settle in villages, to convert to Christianity, and, ultimately, to make deals with development. The clans had historically engaged in internecine warfare, murdering each other at an unsustainable rate and spearing intruders with equal violence. While most of the Huaorani people settled along rivers in small villages, some clans wanted nothing to do with the outside world and still wander the Yasuni forests to this day as their ancestors did.

Joe Kane in his book *Savages* describes the struggle the Huaorani have had with the intrusion of big oil and logging in the north Amazon Basin. They are still trying to make Chevron pay the billions owed them from an international court ruling for desecration of land, water, and the health of their people.

Kemperi, a shaman in the village of Bameno, had a message to the foreign peoples who live where the oil companies come from:

My message is that we are living here. We are living bien [in a good way]. No more [oil] companies should come, because already there are enough. They need to know that we have problems. . . . Many companies want to enter, everywhere. But they do not help; they have come to damage the forest. Instead of going hunting, they cut down trees to make paths. Instead of caring for [the forest], they destroy. Where the company lives, it smells nasty; the animals hide; and when the river rises, the manioc and plantain in the low areas have problems. We respect the environment where we live. We like the tourists because they come, and go away. When the company comes, it does not want to leave. Now [the company] is in the habit of offering many things; it says that it comes to do business, but then it makes itself into the owner. Where the company has left its environment, we cannot return. It stays bad. Something must remain for us. Without territory, we cannot live. If they destroy everything where will we live? We do not want more companies, or more roads. We want to live like Huaorani, we want others to respect our culture.

The Ecuadorian government, led by President Rafael Correa, who billed himself as an environmental leader, kept opening up more plots for oil and mining bids, mostly to China in exchange for a $7 billion debt. The fight moved to the pristine southern tier of the basin and engaged the huge tribe of Shuar, numbering close to fifty thousand including their Achuar relatives, about six thousand, to the south and east on the Peruvian border. Their joint lands extend for hundreds of miles and their past as headhunters is well known. For the Shuar on the Cordillero del Cóndor, a lush rainforest on the Andean slopes, the arrival of drills for a Chinese gold mine was tantamount to a declaration of war. They began sharpening their spears along with their Achuar allies. The Shuar chief Domingo Ankuash said, "The strategy is to unite the Shuar like the fingers of a fist. The forest has always given us everything we need, and we are planning to defend it, as our ancestors would, with the strength of the spear. To get the gold they will have to kill every one of us first."

This was what Sumpah was referencing when he spoke of the need to come together to fight for nature. Fernando told us that although they had not shrunk human heads lately, they kept the practice alive by shrinking monkey heads. I shivered a bit when we rose to leave Julie with her Achuar shaman for the next eighteen hours. As we motored upriver, Julie's waving figure on the red cliff became smaller and smaller until she disappeared entirely, and I wondered what on earth I was thinking bringing this dear Nova Scotia friend to this remote and volatile part of the earth.

An encounter in the forest calmed me. There was only peace there. A gigantic kapok tree, the sacred ceiba of Mexico, two hundred feet tall, twenty-five feet in girth, its anaconda-like root-arms undulating across the forest floor, was where the spirit seekers came at night and where they received Arutam. I could feel it. The tree had been alive for hundreds of years and sheltered many a soul, a thing of great beauty and history. Within its crevices lived frogs and insects; bromeliads climbed on its vines. Its seedpods had probably floated across the Atlantic eons ago to colonize Africa.

Later that night the Norwegians and I drifted in the canoe, as the endangered Black Caiman, hunted for its exquisite crocodile leather, reflected blood-red eyes on top of the black water. Bats flitted over the surface, ridding the air of insects. One the size of a little butterfly hovered in front of my nose as if deciding I might be a meal. It could have been a tiny Vampire Bat; there were many species unknown to me. A Ladder-tailed Nightjar sat motionless on a stump pretending it couldn't be seen in our flashlight's glare, while an Agouti, or "Paca," ran under the boardwalk, and an Armadillo scurried into the brush.

Julie arrived back before breakfast full of enthusiasm for the ayahuasca vine and the ceremony in Sumpah's village. She and others, including Fernando and Simon, had lain on large banana leaves under the stars while Sumpah spoke of what was to come and the drink was presented in bowls. It was several hours before the visions began, hours of purging the ayahuasca from her body before drinking more and lying back down to Sumpah's endless chanting. Then she felt the spirit come to her. A gray wildcat with stripes crawled up and lay on her chest purring, and another of the sacred animals, an Anaconda, curled around her head and body. She was not afraid. She lay there peacefully listening to the night, to babies coughing in their sleep and mothers soothing them. Sumpah interpreted Julie's visions. He said it was a very good sign to have more than one spirit animal visit. Soon the horizon brightened and the little village came to life.

Julie experienced a profound change. A year later she told me it had been an "awakening," an "opening up to a new reality . . . my sense of being separate from everything has been removed like a cover over my eyes and I see that . . . what is in you is in me . . . that I don't live life but life lives through me."

To the four basic elements of earth, air, fire, and water, a fifth is sometimes added, an invisible element. Aristotle called it aether, Hinduism akasha, or that which was beyond the material world, and Buddhism recognized it as an experience of the senses. In science it has the power to move things. Many tribal people in the world today who live with nature keep the invisible spirit life whole and within them. It encompasses a vast cosmology of waking life and dream life and embraces all things seen and unseen in nature.

In our metal and stone cities with hard edges and constant contact we do not connect often with this spirit world. We have our churches, our music and art, and our love for each other, but the wonder of the natural world is often lost to us. God goes by many names. Whether we call on Zeus, Arutam, Shiva, Jehovah, or the Goddess, there are no atheists in wilderness.

23

Ocean

Paleontologists are unsure which land mammal related to whales entered the water fifty million years ago and never left, becoming the largest animal ever to inhabit the planet, including dinosaurs. It excites me to think we still share Earth with the Blue Whale, 100 feet long and 160 tons. Although whaling took them to the brink of extinction, with almost 350,000 killed in Antarctica alone, by 1971 whaling of these giants ended and the Blues are now slowly rebounding off California and the eastern North Pacific, where more than 2,200 were counted in 2014. Colliding with boats may be what they have most to fear today.

My childhood summers growing up on Nantucket were filled with whale talk, whale bones, and whale paraphernalia. We spent rainy days in the Whaling Museum poring over ships' logs, the same ones Herman Melville read before he wrote his great novel. The hair-raising tale of the ship *Essex* stove in by an angry whale in the South Pacific became the basis for *Moby-Dick*.

In the fifty years I spent on Nantucket I never saw a whale offshore, although my mother and I would comb the horizon. We took trips out of Cape Cod Bay in tourist boats to see pods of Humpbacks identified by name because of the singular patterns on their flukes: Salt, Anchor, and Ganesh can still be seen. My love affair with whales continues to this day.

In Hawaii, Humpbacks sang as I dove under the waves. I became transfixed by the soulful keening, answered from the deep in whale chorus. How remarkable that the sound penetrated through the heavy water!

Humpbacks seem to enjoy human company. Ed and I went out in a Zodiac off Maui, a twelve-foot yellow blow-up captained by a sun-kissed young man. We soon found ourselves surrounded by a pod of almost twenty Humpbacks, a rowdy bunch of bachelors that began circling us, diving under the little boat, gently lifting it a few inches, and then breaching eighty feet away, their entire bodies rising out of the water in unrestrained leaps like victorious football fans at touchdown. Our young guide was over the moon with excitement and begged us to jump in the water with them. "They want to play!" he kept exclaiming. Alas, and I still regret it, I didn't have the courage to do so.

The most affecting story I heard about Humpbacks was from my friend Victor Perera. In Northern California a whale became entangled in fishing lines, which threatened to drown it. Divers worked for hours to free the whale, and when the last line was removed, rather than swim away the Humpback gently bumped each man, the great whale eye making contact with the human one before returning to the deep.

On pelagic trips when I go out to sea with birding buddies to sight shearwaters, skuas, and jaegers, we sometimes glimpse Humpbacks or Finbacks breaching in the late-day sun on the Atlantic horizon. Or the rare Bottlenose Whale, its melon head and smiley beak popping up to have a look at us in the Gully, an ocean trench 125 miles off the coast of Nova Scotia. The Bottlenose and other beaked whales are particularly sensitive to mid-frequency sonar, which interferes with their echolocation. They will move rapidly away from military sonar testing in the ocean, but not fast enough in many cases. Thousands of beaked whales have beached themselves worldwide, the victims of decompression sickness and rapid surfacing. Joshua Horwitz recounts the long fight to end sonar testing in his riveting book *War of the Whales*.

Along the California coast on a pelagic tour of Monterey Bay with the legendary seabird expert Debi Shearwater, fellow birders and I watched seven Orcas, a pudgy black and white baby safe in the middle of the pod, chase Gray Whales as they migrated north, hoping to snatch one of *their* babies. The mother and nurse Grays kept their young protectively on the starboard side nearest the shore as they cruised their way to Alaska.

When Ed and I moved to Nova Scotia I was thrilled to see Pilot Whales and occasional Minkes off our shore along with Porpoises and Dolphins. But the oddest encounter was with the carcass of a Finback, the second largest and my favorite whale. They barely unveil at the surface, their single fin cresting out for only a moment like a periscope before submerging again.

Roger Payne has studied whales most of his life. He found that Fin Whales emit deep-frequency sound waves traveling special ocean channels for thousands of miles where the speed of sound is slowest. On quiet days when these channels are not disturbed by the noise of ship traffic, the Finback's sounds can travel as far as thirteen thousand miles, half the diameter of the earth! These channels are also used for submarine warfare, and before it was discovered that the Fin Whales made these sounds the government thought they were from foreign military sonar.

There is a small cove, about one hundred yards of beach, just south of our place on the Nova Scotia coast. The ocean waters beyond it are fairly shallow, ten to eighty feet deep, with numerous granite islands jutting up, good nesting sites for Eiders, Black Guillemot, Cormorants, and some Puffins, but tough to maneuver for both big boats and big mammals like whales. Historically this part of the south shore was known as the Ragged Islands.

A fifty-five-foot Fin Whale washed ashore in this small cove, and died of unknown causes. Because the weather was warm and the stench severe, the whale was buried deep in the sand. A year and some months later between Christmas and New Year's, Ed and I found another Fin Whale, sixty-five feet long, dead in exactly the same spot. No one bothered to bury this one because it was so cold.

We watched over the course of a week as Herring and Black-backed Gulls first took the eyes, then made their way through the baleen to eat some of the huge tongue, and then pecked down the back, stripping the skin along the vertebrae to get at the blubber. Other animals must have come in at night, perhaps Foxes, Coyotes, or Great Horned Owls, because the flesh was almost gone in ten days.

It is a mystery why the great whale came ashore in exactly the same place. It is not easy to negotiate the Ragged Islands. Were the two related? Like Elephants, are they able to find their kin? Is this cove a whale graveyard? These questions remain unanswered, as do many questions about ocean life.

Robert Ballard, director of the Ocean Exploration Trust, said, "We played golf on the moon before we went to the mid-Atlantic ridge, which is the single largest geological feature of our own planet; we have better maps of Mars than of some parts of our ocean floor." The variety of life and life systems—the biodiversity of the ocean—is staggering. It is estimated that 50 percent to 80 percent of all life is under the sea, and 95 percent of it remains unexplored. Too bad young Charles Darwin didn't get the chance to jump from the *Beagle* into the waters of the South Pacific. He would have been gobsmacked. It is really only now, however, that we have technology advanced enough to penetrate to the bottom of the ocean floor and shine a light on the creatures there. Most of them have their own luminescence, as sunlight does not penetrate below three thousand feet. They are stars in their own firmament of darkness.

The ocean is full of man-made perils. The rise in acidity due to carbon fallout threatens diatoms, corals, fish, and people and will not be able to sustain many forms of life for long. And as the trash bin of the world, the ocean seemed limitless. We thought it could absorb whatever we threw at it: garbage, oil, chemicals, radioactive containers, or plastics. Huge gyres of waste circling in ten-mile-wide whirlpools are now exposed through satellite photos. Nothing can hide anymore. Fish the world over are found to be contaminated with mercury, tiny beads of plastic, and the general effluent of human trash.

The Northern Gannet, my favorite seabird, plunge-dives from more than fifty feet straight into the ocean for baitfish. It is stunning to see this large white bird streamline its wings against its body like an Olympic swimmer and split the water with its long gray beak. Every summer I find a few dead Gannets on our beach. I cut one open and found bits of blue plastic, probably a fisherman's recycling bag, in its stomach. He must have mistaken its glint in the sun for small herring.

Commercial fishing vessels have exploited many coastal areas, exhausting legal limits. The deep ocean, two-thirds of which lies beyond the boundaries of any one country and is a free-for-all without regulations (although the UN is trying to impose them), is where fleets are fishing now with high-tech gear, staying at sea for years at a time like the old whaling vessels did. This puts world panocean travelers like the Bluefin Tuna in peril.

Carl Safina, one of our finest nature writers, has three pages in his elegiac book *Song for the Blue Ocean* describing why the Bluefin is so remarkable:

> The bluefin's immense strength and stamina are not mere by-products of its size. Some sharks get big, but the strength of a five-hundred-pound shark does not compare to that of a five-hundred-pound bluefin . . . the combined, coordinated functioning of millions of muscle cells that are among the most powerful and specialized in any creature . . . Making the bluefin so unbelievably tough is a body thoroughly designed to penetrate cold, food-rich waters and rule as the top predator there . . . Of more than thirty thousand fish species plying the world's waters, the bluefin tuna is among the few that have developed the ultimate weapon of vertebrates: heat.

It is heat racing through the vascular network of the Bluefin, the billfishes, and a few sharks like the Great White, the Mako, and the Porbeagle that make them so specialized and unique. There is simply no ocean animal as powerful. No wonder men have been strapped into chairs bolted to slippery decks in an effort to reel in these masterful creatures. It is like taming a giant stallion.

In the 1950s, tuna fishing attracted Hollywood stars, British royalty, and machismos like Ernest Hemingway to the little town of Wedgeport, Nova Scotia. My father and his friend John Morris were among those casting their lot for the great fish. On one trip John caught a whopping 780-pound Tuna that took him eleven hours to reel in, while Dad had one not so niggling at 395 pounds. Its tail was mounted and hung proudly in our living room for years.

Commercial fisheries for the giant also began in the 1950s, but that and sportsfishing did not deplete the population of Bluefin Tuna. Our palates did. With the introduction of sushi and sashimi to gastronomy in the 1960s and '70s the Bluefin had more to fear than the Japanese; it had the entire Western world after it. The species plummeted an estimated 85 percent. By 1999, the International Commission for the Conservation of Atlantic Tunas, or ICCAT, had a twenty-year rebuilding plan that halved catches and placed moratoriums.

Overfishing is the single most egregious factor in the depletion of fish stocks around the globe. While technology has been a boon to fishermen, revealing weather patterns and fish migrations and mapping the ocean floor, it has been a disaster for fish, which used to have a sporting chance. It has also been a disaster for seabirds, turtles, dolphins, and whatever else is caught by accident in nets or on hooks.

In the South Pacific, albatross and petrels dive for baitfish on one of thousands of hooks trailing sometimes thirty miles behind huge long-lining vessels. They get hooked, pulled under, and drowned. The Wandering Albatross, with the largest wingspan of any bird at more than eleven feet, is a master of aerodynamics, riding a current of air over the waves, rarely flapping at all. It spends its life at sea except for breeding season, and then the pair, mated for life, return to the same remote island and lay a single egg in a colony of thousands.

Wisdom is the name of one Layson Albatross, the oldest living banded bird on earth. She was banded on Midway Atoll in 1956 and was thought to be sixty-five in 2016. She and her mate have raised thirty to thirty-five chicks through the years, which they incubated and fed for seven months before they fledged. She still lays an egg a year. She flies about fifty thousand miles annually in search of food, putting her at the top of the million-mile club.

Nineteen of the twenty-one species of albatross are in trouble.

They get hooked on long-lines or they ingest plastics and other litter, or they are poisoned by chemical runoff from ships and coastal sites, or rats and snakes eat their eggs and chicks.

Most fishermen do not wish to harm bycatch they hook or net. They don't want to be fined, either. If it is economically feasible for them they will switch to nets that release turtles and other creatures, they will tie ribbons on their lines, and they will use better hooks. More and more fishing vessels are adopting new technology, such as that invented by a conservationist and two Ecuadorean fishermen. This device not only saves birds, it saves time baiting hooks and injury to fishermen's hands. Hundreds of baited hooks on lines are chuted into a six-foot tube that then goes overboard, quickly releasing the hooks underwater before the birds can grab them. This kind of ingenuity has saved many an albatross. There will be more smart inventions and more regulations in place in the future. Many people are working on it. Technology is both the nemesis and the savior of conservation efforts.

Salmon is a fish that feeds the world, and we have done a lousy job of keeping it wild and healthy. It is a gorgeous creature, revered by Native Americans for thousands of years. It has a complex life cycle that takes it to the open ocean as a smolt and back to the same river, sometimes the very place where it was hatched, one or two years later. The female lays her eggs in the clear gravelly bed of a pool, the male deposits his milt over them, and the process begins again. The old salmon die in these streams, becoming nutrients for tiny aquatic species as well as bears and eagles.

Because of pollution, damming, and silting of our great rivers, salmon species have declined so dramatically that Alaska is the only place in North America where great runs occur consistently and the stocks have been well managed.

Most salmon is farmed today. Aquaculture is controversial because the business of feeding and growing salmon in ocean pens is a dirty business. Keeping salmon in pens compromises the very heart of these long-distance migrants. They are carnivores and are fed pellets of ground-up baitfish like mullet, herring, and sardines, compounding the depletion of smaller species down the food chain.

In southwestern Nova Scotia there have been salmon wars over the farmed fish pens dotting the mouths of rivers, harbors, and estuaries. This is lobster country, and the fish pens are not clean. Fecal matter, waste food, and chemicals drift through the cages to the ocean floor and accumulate in piles many feet thick that disperse with the currents to smother seaweeds, shorelines, and larvae.

I was part of a breeding bird survey offshore in the boat of a fisherman who had been hired to dive down to check sediment levels for an aquaculture company. He described it as a horrific dead zone radiating far beyond the pens. Lobsters would not go near it. These blights on the ocean floor are as irreparable as mining sites above ground.

Salmon are given antibiotics to ward off infections that become rampant in crowded pens. Infectious salmon anemia (ISA), sea lice, and worms invade their flesh, although we consumers are told they are not harmful to humans. In addition, the salmon are fed pellets with minerals that contain copper, and copper is also found in the chemicals used for cleaning the nets; this is an element that does not break down, meaning the water remains contaminated in perpetuity, threatening all living things, including bathers on the beach.

The salmon wars have become ugly, with neighbor pitted against neighbor. Signs sprout in front of homes for and against aquaculture, and fights have broken out at meetings. Southwest Nova Scotia is billed as the "Lobster Capital of Canada." Spawning grounds are in front of our house; if you dive down into the water you can see mother lobsters waving their claws in defense of their nest bowls

Citizen action in southwest Nova Scotia in the summer of 2013

on the clear sandy bottom. The chemicals fed open-pen salmon are
toxic to lobster larvae, as well as adult lobsters.

I eat farmed salmon. Not often, but I have little choice in rural
Nova Scotia where the last wild salmon barely cling to existence in
a few rivers. Acid rain from industrial sites in America's Midwest
has been blowing in from Indiana or Detroit on prevailing westerly
winds for decades, destroying the delicate pH balance that makes
eggs viable for salmon and trout. Enormous efforts to keep wild fish
alive are now compromised by the added problem of escaped farm
salmon that interbreed and dominate the gene pool, weakening the
wild strain, in addition to introducing disease. Fly fishermen are
staring at the extinction of the wild Atlantic Salmon.

Nova Scotia's government gives big subsidies to aquaculture corporations, knowing that farmed fish is future protein for ten billion
people—unless the world takes a sharp right turn and embraces
vegetarianism. But many of the corporations mismanage their farms
and take no responsibility for cleaning up the mess they make in the
ocean. The government either has no regulations or does not do a
good job enforcing them. Too often the entire stock must be killed
because of disease, further fouling the waters. Here is a plea from a
local resident on behalf of lobstermen, families, and all of us:

The removal of dead and dying fish from 3 open pen aqua-culture sites in Jordan Bay and Shelburne Harbour, NS began last week and is ongoing. Thousands of gallons of waste water and debris are being released in both bays every day. Rocks are coated with thick grease. The beaches and salt marshes are strewn with salmon parts and grease. Pieces of salmon are floating in the water and salmon carcasses in various states of decomposition are washing ashore.

This type of activity is permitted by the provincial gov-ernment, in lobster and traditional fishing areas. Private prop-erty and recreational activities are also impacted. The Director of Aquaculture told an upset local resident this week that "this is an efficient means of waste disposal."

An obvious question would be, why is this acceptable any-where in coastal/public waters?

Why indeed? Why is Big Oil permitted to drill anywhere it has demonstrated it cannot handle spills? How long will the public's health be compromised before big corporations take responsibility for the damage they inflict on all living things? Why does an acci-dent have to happen first before a company or a government takes responsibility for outcomes?

The old fish-processing plant in Clark's Harbour, at the southernmost point of Nova Scotia and Canada, has been totally refitted as a Halibut nursery. Shelley LeBlanc met me outside, a tall gal with porcelain skin and a model's figure. She picked up a cozy black cat at her feet, Rocky by name, and we went inside. Scotian Halibut has been around since 2001. Halibut take a long time to mature and the females are not ready to breed until they are seven or eight years old, the males a bit younger. Local fishermen brought their initial mature Scotian Hali-but adults from deep waters, and they soon had their first spawn.

Iceland and Norway are the champs in this field. Iceland has been farming Halibut since 1980. They co-invested with Scotian Halibut and lately have been getting stock from them as their own fish stock collapsed; they are not saying why, but disease is usually the culprit. There are ten hatcheries in the world; Scotian Halibut is the only one in North America, and Shelley is the only person in North America doing what she does.

Halibut—*Hippoglossus hippoglossus*—are deep-ocean flatfish of the Atlantic from Norway to Labrador, Nova Scotia, and the Gulf of Maine. They are big bodied, sometimes reaching six hundred or seven hundred pounds—or they used to. They have been overfished and have been on the IUCN's endangered species list since 1996.

Shelley led me into a cold, dark concrete room on the ground floor where the temperature was about forty-five degrees. There were rows of large tanks the size of aboveground swimming pools with water running through them. She shone a flashlight on one, and there were the most beautiful little translucent eggs hanging in the water world like tiny galaxies. They eat nothing at this stage; they just float there for two weeks in forty-three-degree water. She says making them is easy. You simply mix male milt and lady Halibut eggs in salt water and poof—the eggs develop. On average a female will give about one-half to a full liter, or forty thousand eggs, in one spawn. The champion ever recorded anywhere was a two-hundred-pound female that gave 2,182,773 eggs. In the wild, Halibut eggs drift suspended thirty to fifty fathoms down, feeding many different ocean dwellers.

There were about fifty thousand eggs in this one incubating tank. They are moved into another tank in another room for the longest yolk sac period of any fish—about six weeks. When they hatch and are around an eighth of an inch long they are given Artemia, a tiny brine shrimp that looks like a winged bird under the microscope. The tanks are vacuumed daily to remove clay (which is added for turbidity so the little fish do not go toward the light all the time), dead fish, and Artemia.

When they are half an inch long they have eyes on either side of their head like any regular fish; their bodies are translucent and the Artemia is orange so you can see it moving through the tiny loops of their intestine. The survival rate is about 50 to 60 percent. At

forty to fifty days old they become flatfish and their eyes move to the
right side of their heads, although there are occasional lefties. Many
don't survive past this stage, and getting five thousand is a success.
The last tank room held these precious survivors, which were now a
few inches long and looked like miniature Halibut, mottled brown
and white and gray. Massed together at the bottom of the tank, they
seemed like an intricate mosaic. In the wild, Shelley said, it is very
hard to see them. They come to shallow inshore waters to grow, their
coloration blending in with sand and rock so beautifully that they are
often overlooked by prey and predator alike. In the tank they feed
on pellets of fish and algae-based lipids.

This is when they are shipped to Norway, Iceland, Prince Edward
Island, Hawaii, and other places in the world for more growing and
to market, a process that takes three to four years. One thousand of
them can fit in a container box about four feet by six feet with cold
water and oxygen for the forty-hour journey by FedEx.

In the last room, a gigantic swimming tank dominated, and we
climbed up to a platform. Thirty-five big Halibut, most of them
brood females, one weighing two hundred pounds, circled the tank.
When they saw Shelley, they congregated like eager heifers to greet
her. There were a few smaller males in there to increase the phero-
mones. Shelley called them all by name and they lifted their big lips
out of the water for the frozen herring in her hand. Lucy was one
of her favorites, and Meryl and Saphoria were the biggest. I tried to
feed them but became shy when she warned me their teeth were very
sharp. I touched one; her skin was as soft as a baby's bottom.

A few months later, when Shelley had tricked them into thinking
it was time to spawn, through light and temperature manipulation, I
watched from the platform as she and her husband, Dan—another
"fish person," she said proudly—slipped into the water in dry suits.
She and Dan have no children—they call these Halibut their babies.

They wandered the tank with penlights looking for two particular
females. When Shelley found one she nudged it with her foot and it
surfaced enough for her to lift it from underneath and slide it onto
a table. Her hands were bare so as not to harm their skin; she keeps

a pitcher of warm water on the table so she can warm up her fingers between fish. Shelley held her down gently as the fish thrashed, searching for gill breath in this element, her tail curling up and back in panic. When she finally calmed down, Shelley stroked her lower belly under the gill with the back of her hand pressing slightly to force the eggs out. Dan held a container and eggs poured forth like tapioca from a fountain. The first lady gave almost half a liter and the second the same. It lasted only a few minutes before they were lifted back into the water and swam away. These are gentle creatures, like cows, Shelley said—cows of the ocean. She called them her "girls." Some brushed up against her in an affectionate way. Hannah was one of these, a big girl who was out to pasture because she was not spawning anymore. Shelley is passionate and humane; she will not let her girls be culled when they don't produce anymore and lets them live out their lives either in the tank or in an aquarium. The Ripley's Aquarium in Toronto is where two of her girls happily reside. Hannah glided by in the water gazing at Shelley adoringly with her opaque blue eyes comically close together on the right side of her head.

Before they left the tank Dan and Shelley took the milt from two males. It poured out of a similar orifice near the gills like thick milk gushing from a spout. Shelley set it aside to be cryofreezed and took four vials of already-frozen milt and poured them into the containers of freshly stripped eggs. Then she told me to gently swish them together with my hand in the frigid salt water. Within a minute, sperm found egg. I found it very moving to be the catalyst for thousands of fertilized eggs with one swirl of my hand. We took them to the first room, releasing them into their own huge dark tank to begin the process of incubation all over again.

Shelley is special. How many people in the world can say they raise baby Halibut for a living ?

The sun hit small translucent marbles on the beach as far as the eye could see, sparkling like a crystal road. What was this tumbling in with every wave all afternoon? They rode the crest of the rollers and spilled onto the sand and rocks at the end of our town's Crescent Beach, sometimes piling four inches high.

They were not hard at all. They were like jelly, glass jelly, like something Salvador Dalí might have painted, rearranging expectations in your mind. We all know Moon Jellyfish in this part of the world, and the largest jelly of all, the Lion's Mane, with their orange fuzz and tentacles. These were different. They were crystal clear and inside was a perfect little indigo body like a tadpole, a tiny oblong of the brightest blue caught inside a glass globe.

Migrating birds went crazy in the September afternoon. Stoking up for the long flight to the Caribbean or South America, Sanderlings and Semipalmated Plovers dove their beaks into the gelatinous mass in a frenzy, but only the Sanderlings stayed. The others decided it was not for them. The Semipals rubbed the jelly off their beaks onto their breast feathers in an effort to extinguish the taste.

"They're tunicates," said a passerby.

"Tunicates?" I replied.

"That's what someone said earlier. Google them."

All the beaches were covered with them, a mile in each direction. It was a massive invasion. I collected a jarful to photograph; before I finished shooting them, arranged in clamshells and tiny pools of seawater, the jelly had disintegrated and the little blue creature was losing color.

There were thousands of different kinds of tunicates on the Internet. There were photographs of ones called sea squirts, some as big as a chair, attached to the ocean floor or rocks in a variety of shapes and vibrant colors. I saw none that looked like our little beauties.

I contacted Nova Scotia's marine biologist Boris Worm, who told me that, indeed, they were tunicates and that they were of the phyllum Chordata just as we human beings are. They have a "notochord," a kind of backbone or rod, which is found in all embryos of Chordata. I find this amazing. We are connected. These beautiful creatures were not jellies at all. They had a spine, a nervous system; they had a complex gut, and sexual and asexual ways of reproducing. They live inside their translucent globe for filtration, bringing water and phytoplankton in to eat and excreting fecal matter out the other end.

They are very, very old, first appearing around 500 million years ago. And they are common in the oceans of the world. Larry Madin of the Woods Hole Oceanographic Institute is one of the world's experts on gelatinous marine animals. He thanked me for sending the pictures and said they were of the order Salpida, or Salps, and the species *Thalia democratica,* a small tunicate that does not attach itself to rocks but spends its life free-floating in the deep ocean or the slopes of the continental shelf. At night they descend to two thousand feet but come close to the surface during the day. They can form long chains that move together through the water and reproduce asexually, or they can roam the water as single sexual beings. They are food for countless species, from large baleen whales to sea squirts. They also eat carbon-producing algae and so may mitigate against emissions that increase global warming.

Sometimes they swarm, creating huge mats that are carried by ocean currents to beaches such as ours. They are harmless and soon die on the shore, lost to the hungry mites of the sand that devour the dead. We human beings are made of seawater and stardust. These Salps share that with us.

The fog was dense. There were about forty of us on the deck, a few yellow slickers piercing the wall of white. The weather was not unusual in the Bay of Fundy; still, it came on quickly, blanketing the water, making any whale sighting impossible. The captain cut the motor and told us to be very still. From port to starboard we waited, not moving a muscle. The lap of waves against the hull was gentle, and we moved with the cradle of the boat. We stared into the mist with the expectation of travelers in a brave new world. How long, I cannot remember. The whale surfaced without our knowing and then we heard the huge animal draw breath, a deep whoosh of air filling the great lungs before it slipped home to the dark waters. The breath of life.

EPILOGUE

We live in the most extraordinary time: massive changes to our climate are juxtaposed against the continuing miracle of the human brain, which recently transported us to see pictures of the farthest dwarf planet in our solar system. And we are connected in milliseconds to anyone in our world through handheld devices transporting our images, sounds, and ideas through the ether.

This instant connectivity has made it impossible to hide anymore. We have no secrets, there are no uncharted territories, and there is nowhere without the sound of human activity. A few minutes of silence snatched here and there in remote places is sure to be interrupted by the buzz of a plane or a saw or a phone. We are transparent and stand naked on the threshold of the future. And for all the exploration of space, all the cold stones and ice, there is no planet B. Earth is our home.

It is only through thoughtful management that we will save the declining species of the world. Tom Lovejoy said it many years ago: "We still tend to think in the very short term and locally when in fact we are disturbing global systems and the way the planet actually works. We need to consciously manage the planet." We will not save all species. Those that need very particular niches to survive may not make it. But there is life everywhere, and life *wants* to thrive. Life-forms have been at it for millions of years. Birds, fish, and some mammals continue their ancient migratory patterns despite the buffeting of changing weather and human incursion. Many adapt. We can help them when they arrive in our area.

Our lines of connection have also created extensive communities of activism. Information is shared online, petitions are signed, and legislators are paying attention more than ever. There are thousands of organizations dedicated to saving the environment. This was not the case fifty years ago when I joined the Sierra Club, the Audubon Society, and a handful of others. Now we can surf the Internet and find organizations dedicated to saving bats, butterflies, sloths, orchids, or bogs. We can "adopt" an Elephant or a gorilla or a tortoise, or pledge to save habitats ensuring clean air, water, and landscapes.

We do what we can in the time of our lives to hold things in place. It is vital to do so. I believe it is a moral imperative as the most evolved creature on the planet to care for the home we share with all others. Everything we need, or make, comes from natural resources: food, clothing, shelter, minerals for technology. The rest is a constant dance with our imaginations, retelling stories a thousand ways and conjuring numbers to understand how the universe works. Without wild things and wild places our stories are diminished and we ourselves become stunted.

I cherish the hours I have spent with scientists in the field. I admire their patience, perseverance, and endless wonder at the natural world. In the face of so much bad news I felt sure they must be pessimistic about the future. But WCS's chief of conservation, John Robinson, reminded me that you cannot be a conservationist and a pessimist. It is an oxymoron. The endeavor of conserving anything is an effort of optimism and hope.

There are almost fifty million people in the United States who call themselves bird watchers. Some will venture outside of their own region as I have done, some will visit nature centers and take part in bird counts. Most are content to watch birds and animals right where they live, marveling that a bit of unpredictable wildness visits them.

I am one of those people. Of all the places I have been, my own backyard, my "patch," is the place I know and love best. It is deep in my heart: the soil and the scrub, the stream and the tide; it is where I watch the sky for birds as the seasons change, those winging in during spring to rest before heading north and those coming home to breed. My patch begins around my house and extends through the neighborhood and beyond as I learn the ways of creatures in nearby places. Protecting those species is my patch work.

Bon Portage Island is close to us. It lies a mile over the water, off Nova Scotia's south shore, just an hour away. Fifty thousand Leach's Petrels nest on this foggy outpost of windswept spruce trees. There were burrows everywhere on the forest floor, little holes peeking between tree roots and against spruce logs, making every step on the sod a bit precarious. I gingerly placed a toe down, testing the sphagnum to make sure I was not going to crash through the roof of a petrel home.

The young biologist gently cupped and drew a tiny Petrel chick from a three-foot burrow beneath the spongy moss at our feet, knowing its parents were off searching for food. They left before first light and would not be back until dark, traveling over the ocean perhaps a hundred miles to find tiny fish for their chick. She assured us they would not mind our touching their baby. I cradled the precious chick in my hand as it bleated in high dudgeon. Gray fluff radiated from two black eyes and its smell was deeply sweet, like violets in loam after a rain.

Bon Portage is owned by Acadia University and managed by the Nova Scotia Nature Trust. Students come to research the bird life and band some of the migrants passing through on their way south in the fall and north in the spring—thousands of birds, from warblers to thrushes. It is a protected island, considered one of the most important habitats in all of Nova Scotia, if not the northeast Atlantic. The protection offered by the university gives hope that the island will be home to these tiny birds forever.

Well, not forever. There is no "forever." Things change. Things fall apart. New things come into being. The little petrels' burrows on

Bon Portage Island will almost certainly be inundated by rising tides, which are already apparent on Nova Scotia's coast as the Greenland ice sheet melts and sweeps into the Labrador Current. Not this year or next, but some time. For now it is safe.

My patch is a kind of island, although connected to my neighbor, and she to the next, and the next and on and on; our quilt can blanket the world. I protect my patch for the creatures who live there. I don't poison them or their grasses or trees. I encourage the plants they like to eat and nest in, and the nectar they sip from flowers. In return I am serenaded with song in the morning, the Muskrat grazing on pond reeds, and the eels making circles in the water at night to stir up insects. I welcome the sweet trill of the Yellow Warbler in the wild rose, and the Black-throated Green Warbler in the woods.

As the days grow short, shorebirds return from their tundra nests and head down the coast, stopping for worms and insects in the seaweed on our beach: Black-bellied Plovers and Least Sandpipers and Ruddy Turnstones. The Blackpoll Warbler makes the longest flight of any small songbird, 1,700 miles over the Atlantic Ocean heading for South America. They visit for only a few days, but I marvel when they stop on our spruce trees before launching themselves headlong on their epic journey.

The Short-tailed Weasels scurry through the scrub into the barn for the cool nights while the big sow Porcupine leads her waddling youngster up a tree, her inquisitive black eyes peering at me, sensing I am no threat. In her seasonal quest for sodium she has overturned the mossy lawn looking for larvae but it is okay with me. We are in this together.

Acknowledgments

Nothing happens without others. We are social animals. I am first indebted to Delauné Michel, who introduced me to her literary agent, Laura Yorke, at the Carol Mann Agency, the perfect home for me. Laura and Carol directed me to interested publishers and Seth Godin advised me to go with senior editor Victoria Wilson at Knopf. Vicky was outstanding from the very first meeting, sensing I had more to tell than the proposal she insouciantly tossed on her desk. This was a fine proposal that Karen Kelly helped me structure, but when Vicky changed its direction I knew I had to go it alone. Her guidance these past two years has been constant and wise. My thanks also go to her able assistant, Ryan Smernoff.

My time with scientists in the field and with colleagues in board meetings or at conservation events always ended up in my trip journals, making it easier to recall conversations and sightings. First and foremost, I am grateful to my friend Alan Rabinowitz, the Tiger Man in my life, who tapped the conservationist in me and introduced me to a home at the Wildlife Conservation Society. William Conway, Archie Carr III, Elizabeth Bennett, John Calvelli, John Robinson, Edith McBean, Joyce Moss, John Gwynne, and the inestimable George Schaller taught me more than they will ever know in those years.

The Audubon Society has been a constant in my life. The first chapter I joined was Hudson Highlands in Putnam County, New York, where I came under the tutelage of the late Tom Morgan, the indomitable Ralph O'Dell, and the steadfast Henry Turner. I joyfully bird on the Christmas Bird Counts with Eric Lind, Max Garfinkle, and the McIntyre brothers and son, David, Lawrence, and Mark. Times in the forests on CBCs or on Bird-a-Thons are some of the happiest of my life. And I thank Charlie Roberto for making events happen and teaching us all to pass on what we know to the younger generation.

When I became a National Audubon board member, CEO David Yarnold energized all of us with the understanding that where birds thrive, people prosper. We need to roll up our sleeves and get to work addressing climate change, bird-friendly backyards, and the education of our youth. I am particularly grateful for the personal contributions to this book by staff members Chandra Taylor Smith, David Ringer, Peg Olsen, Mark Jannot, Kimberly Keller, Gary Langham, John Beavers, and John Myers.

In Nova Scotia, my birding gurus include Sylvia Fullerton, the late David Young, Ian McLaren, Eric Mills, David Currie, Ted D'Eon, and Ron and Alix D'Entremont. Led by Sue Abbott of Bird Studies Canada, Christine Curry and I have spent many happy hours guarding Piping Plovers on the south shore, while Julie Balish and I have combed beaches and woods for natural objects for her art and birds for my heart. Other friends in Nova Scotia who have been unstinting in their spiritual support include Sonja and Leo Fourdraine, Mike Balish, Cheryl Graul, and Janice Fiske and her husband, Dean, who first took me birding there when he was not at sea harvesting shell-fish.

I am grateful to Ethel and John Andrews, who gave me a room in their home in one of Newfoundland's last outports, and to Cynthia Thomas, who invited Ed and me to visit her in Belize and later in Thailand. Thanks go also to Phil Wallis of Pennsylvania's Audubon Society, who made it possible for me to see the extent of fracking in that state, and to the late Victor Perera, for introducing me to Mexico's Lacandon people and Guatemala.

Thomas Kaplan founded Panthera to save the large cats of the world and I gratefully accepted his invitation to be on its Conservation Council. George Fenwick likewise asked me in the 1990s to join the board of the newly formed American Bird Conservancy, which has been a beacon of strategic thinking. Michael Crowther, director of the Indianapolis Zoo, founded the prestigious Indianapolis Prize to honor field biologists, asking me to participate from its inception.

Lucy Waletsky was an extraordinary chairman of the Taconic Region Commission of New York State Parks, Recreation and Historic Preservation when I served as a commissioner. She has made high art of her philanthropy, generously giving to countless worthy organizations and enlightening politicians about protection of the environment.

I salute the artists who tell the stories of the natural world through performance, literature, and visual art. Nick Brandt's art photographs of African mammals command immediate attention in the galleries where they are displayed. Conservation photographer Joan de la Malla captures the story in a single shot. Steve Winter creates stunning portraits of wild animals and his partner Sharon Guynup writes passionately about them, while composer Steve Heitzeg creates music of the spheres.

There would be no book without the scientists who have been good enough to talk with me or to suffer my presence in the field. Russell Mittermeier sent me glorious books including his own huge compendium of lemurs. Patricia Wright showed the lemurs to me in Madagascar. Carl Safina wrote of the vast water world we inhabit and the intelligence of Earth's creatures. Tom Lovejoy's persistent political savvy has changed minds. Bruce Beehler opened up the world of Papua New Guinea to me; his enthusiasm for the Birds-of-Paradise was contagious, as was Lisa Dabek's for her beloved Tree Kangaroos. Tino Accau revealed his native Peru to me, as Hishey Tshering did Bhutan. Shelley LeBlanc and Brian Johnson taught me about farming fish, and epidemiologist Frederica Perera taught me about the harm chemicals can do to our bodies.

The lives of Cougars, Jaguars, and Siberian Tigers were illuminated for me by Howard Quigley, as were Lions by Luke Hunter. George Archibald, a Nova Scotian by birth as I am by heritage, became a lifelong friend. His compassion, his love of all things living, and his indefatigable optimism have been embedded in my soul.

We all come home, if only in our thoughts. My brother, Tom Quigley, and my sister, Pam Stocker, share my memories of our father taking us on camping trips across the country and every New Years' Day making us hike a mountain or big hill to start the year with a sense of accomplishment, and of our mother, who brought the little things in the air, soil, and sea to our attention.

My dearest friends, Jane Milliken, Susan Dowling Griffiths, Amelia Dallenbach, Gretchen Dow Simpson, Clare Weaver, Tina Howe, Hope Cooke, her daughter Hope Leezum, Anne Bell, Bardyl Tirana, and the late Susan Sollins have been there for me for as long as I can remember. Jane read part of the book early on and enthusiastically endorsed it, as did Diane Sullivan and my daughter-in-law, novelist Genévieve Mathis, whose cogent notes were particularly helpful.

My stepson, documentary filmmaker Anthony Sherin, was invaluable with the photographs, while my stepson Jonathan Sherin trekked beaches for shorebirds with me. My son, Jace Alexander, and his wife, Maddie Corman, were there for me when I needed time or a hug.

It is my grandchildren that make me feel most appreciative. Children expect the world of us, and although we are giving them a damaged globe they continue forward with enthusiasm and faith. Evan Sherin-Jones helped me with early research on the book, Hazel Sherin gave me a thumbs-up, Isabelle Alexander and Vita Sherin-Jones made every moment of our trip to Galápagos joyful, and twins Mac and Finn Alexander delight me every day with their discoveries. They will change the world.

Every day my world is changed by the thoughtfulness of my husband, Ed Sherin.

Appendix

The following list of organizations involved with the protection and conservation of wild things and wild places is by no means comprehensive. They happen to be favorites of mine, whose goals, principles, and missions I wholeheartedly support and are quoted below. Subscribing to one or more of these organizations expands an individual's power, knowledge, and efficacy in the world of conservation. Being part of a huge community, making positive changes for Earth and its inhabitants, is deeply rewarding.

ABA: "The American Birding Association inspires all people to enjoy and protect wild birds."

ABC: The American Bird Conservancy "is dedicated to achieving conservation results for birds of the Americas."

Amazon Watch "was founded in 1996 to protect the rainforest and advance the rights of indigenous people in the Amazon Basin."

Audubon Society: The National Audubon Society's mission is "to conserve and restore natural ecosystems, focusing on birds, other wildlife, and their habitats for the benefit of humanity and the earth's biological diversity."

There are 460 local Audubon chapters across the United States with similar missions.

BirdLife International "is the world's largest nature conservation Partnership. Together we are 119 BirdLife Partners worldwide from 117 country/territories. We are driven by our belief that local people working for nature in their own places but connected nationally and internationally through our global Partnership are the key to sustaining all life on this planet. This unique local-to-global approach delivers high impact and long-term conservation for the benefit of nature and people."

Center for Biological Diversity: "We believe that the welfare of human beings is deeply linked to nature—to the existence in our world of a vast diversity of wild animals and plants. Because diversity has intrinsic value, and because its loss impoverishes society, we work to secure a future for all species, great and small, hovering on the brink of extinction. We do so through science, law and creative media, with a focus on protecting the lands, waters and climate that species need to survive. . . . We want those who come after us to inherit a world where the wild is still alive."

Centre ValBio, at Stony Brook University, "was created by Professor Patricia Wright in 2003 to help both indigenous people and the international community better understand the value of conservation in Madagascar and around the world."

CITES: "The Convention on International Trade in Endangered Species of Wild Fauna and Flora is an international agreement between governments. Its aim is to ensure that international trade in specimens of wild animals and plants does not threaten their survival."

Cornell Lab of Ornithology: "We use science to understand the world, to find new ways to make conservation work, and to involve people who share our passion."

David Sheldrick Wildlife Trust "is today the most successful orphan-elephant rescue and rehabilitation program in the world and one of the pioneering conservation organisations for wildlife and habitat protection in East Africa."

Durrell Wildlife Conservation Trust, Wildlife Park, and the Durrell Foundation are leaders in "saving species from extinction."

Earthjustice: "We exist because the earth needs a good lawyer."

Ecojustice: "Canadians on the frontlines of the fight for environmental justice . . . We launch groundbreaking lawsuits that level the playing field so industry interests can't trump those of people and the planet."

Ecology Action Centre: "A society in Nova Scotia that respects and protects nature and provides environmentally and economically sustainable solutions for its citizens."

Friends of the Earth "strives for a more healthy and just world . . . We are one of 75 national member groups of Friends of the Earth International, a global network representing more than two million activists in 73 countries."

Greenpeace: "Our investigations expose environmental crimes and the people, companies and governments that need to be held responsible . . . We have the courage to take action and stand up for our beliefs."

Indianapolis Prize "is the largest individual monetary award given for animal conservation in the world and is one of the ways the Indianapolis Zoo empowers people and communities, both locally and globally, to advance animal conservation."

International Conservation Fund of Canada "conserves nature in the tropics and other priority areas worldwide."

International Crane Foundation "works worldwide to conserve cranes and the ecosystems, watersheds and flyways on which they depend."

IUCN: International Union for Conservation of Nature "is driven by two features today: Global production and consumption patterns are destroying our life support system—nature—at persistent and dangerously high rates." The IUCN Red List of Threatened Species has assessed about eighty thousand species of flora and fauna to date from "Least Concern" to "Critically Endangered."

National Geographic Society "has been inspiring people to care about the planet since 1888."

NRDC: The Natural Resources Defense Council "works on a broad range of issues as we pursue our mission to safeguard the Earth; its people, its plants and animals, and the natural systems on which all life depends. As an institution we have six main priorities: curbing global warming and creating the clean energy future; reviving the world's oceans; defending endangered wildlife and wild places; protecting our health by preventing pollution; ensuring safe and sufficient water; fostering sustainable communities."

Nature Conservancy: "The mission of The Nature Conservancy is to conserve the lands and waters on which all life depends."

Nova Scotia Nature Trust "protects Nova Scotia's outstanding natural legacy through land conservation."

Oceana "is dedicated to protecting and restoring the world's oceans on a global scale."

Panthera: Panthera's "mission is to ensure a future for wild cats and the vast landscapes on which they depend."

Rainforest Trust: Founded by the ornithologist Robert Ridgely, the Rainforest Trust "protects threatened tropical forest and endangered wildlife by partnering with local and community organizations in and around the areas that are being threatened. After we purchase acres of endangered land, we empower local people to help protect it by offering them education, training and employment."

RARE "focuses on bright spots in conservation . . . to turn local change into global impact . . . by inspiring communities to take pride in their natural resources."

Riverkeeper: "Riverkeeper's mission is to protect the environmental, recreational and commercial integrity of the Hudson River and its tributaries, and safeguard the drinking water of nine million New York City and Hudson Valley residents."

Safina Center: The center was founded by Carl Safina, and its mission "is to inspire a deeper connection with nature and the motivation to act. We do that by bridging science, literature and film . . . We prompt people to make better personal choices, support smart policy change and create innovative, practical solutions that advance global conservation efforts."

Save the Elephants: The organization was founded in 1993 by Iain Douglas-Hamilton, and its mission is "to secure a future for elephants and sustain the beauty and ecological integrity of the places they live, to promote man's delight in their intelligence and the diversity of their world, and to develop a tolerant relationship between the two species."

Sierra Club: "Founded by legendary conservationist John Muir in 1892, the Sierra Club is now the largest and most influential grassroots environmental organization—with more than two million members and supporters. Our successes range from protecting millions of acres of wilderness to helping pass the Clean Air Act, Clean Water Act and Endangered Species Act."

Snow Leopard Conservancy: "Ensuring Snow Leopard survival and conserving mountain landscapes by expanding environmental awareness and sharing innovative practices through community stewardship and partnerships."

Traffic: Traffic's mission "is to ensure that trade in wild plants and animals is not a threat to the conservation of nature." It does this in part by "investigating and analysing wildlife trade trends, patterns, impacts and drivers to provide the leading knowledge base on trade in wild animals and plants."

Tree Kangaroo Conservation Program: Founded by Lisa Dabek and housed at Seattle's Woodland Park Zoo, the TKCP "fosters wildlife and habitat conservation and supports community livelihoods in Papua New Guinea."

U.S. Fish and Wildlife Service: The USFWS is a bureau within the Department of the Interior, whose mission is "to work with others to conserve, protect and enhance fish, wildlife and plants and their habitats for the continuing benefit of the American people."

Wild Foundation "advances a reciprocal, balanced relationship between people and nature—our Nature Needs Half vision."

WildAid: WildAid's mission is "to end the illegal wildlife trade in our lifetimes. We envision a world where people no longer buy products such as shark fin, elephant ivory and rhino horn . . . WildAid works to reduce global consumption of wildlife products by persuading consumers and strengthening enforcement . . . When the buying stops the killing can too."

Wildlife Conservation Society: "WCS saves wildlife and wild places worldwide through science, conservation action, education and inspiring people to value nature."

Bibliography

Below is a selective list of books about wild things and wild places. The authors have informed, captivated, and moved me with their writing.

I could also say the same about the myriad poems, novels, nature magazines, field guides, films, photographs, and TV series that enlighten and inspire me daily, but any list would be far too long to include here. Suffice it to say, I never miss issues of *National Geographic, Audubon,* and Cornell Lab's *Living Bird* magazines. The photos alone are worth attention. Online I read *Scientific American, Nature,* and *PLOS One* to see what is going on in the world of science. And I am indebted to all the remarkable men and women who are blogging about fascinating subjects.

Edward Abbey, *Desert Solitaire,* McGraw-Hill, 1968

Lawrence Anthony with Graham Spence, *The Elephant Whisperer,* St. Martin's Press, 2009

Kamal Bawa and Sandesh Kadur, *Himalaya: Mountains of Life,* Ashoka Trust, 2013

Carol Beckwith (photographs) and Tepilit Ole Saitoti (text), *Maasai,* Abrams, 1980

William Beebe, *Galápagos: World's End,* G. P. Putnam's Sons, 1924

Bruce M. Beehler, *Lost Worlds,* Yale University Press, 2008

Gertrude Blom: Bearing Witness, Alex Harris and Margaret Sartor, eds., Center for Documentary Photography, Duke University, 1984

Lester R. Brown, *Plan B 4.0*, Norton, 2009

Rachel Carson, *Silent Spring,* Houghton Mifflin, 1962

Bryan Christy, *The Lizard King,* Twelve, 2008

William Conway*, Act III in Patagonia,* Island Press, 2005

Jacques-Yves Cousteau with Frédéric Dumas, *Silent World*, Harper, 1953

Wade Davis, *The Wayfinders*, House of Anansi Press, Massey lectures, 2009

Jared Diamond, *The Third Chimpanzee*, HarperCollins, 1992

Annie Dillard, *Pilgrim at Tinker Creek*, HarperCollins, 1974

Iain Douglas-Hamilton and Oria Douglas-Hamilton, *Among the Elephants,* Viking Press, 1975

Sylvia Earle, *The World Is Blue,* National Geographic Society, 2009

Richard Ellis, *Tiger Bone and Rhino Horn*, Island Press, 2005

Dian Fossey, *Gorillas in the Mist*, Houghton Mifflin, 1983

Jane Goodall, *In the Shadow of Man*, Houghton Mifflin, 1971

Fred Guterl, *The Fate of the Species*, Bloomsbury, 2012

Phillip Hoose, *Moonbird*, Farrar Straus Giroux, 2012

Joshua Horwitz, *War of the Whales*, Simon & Schuster, 2014

Joe Kane, *Savages*, Vintage, 1996

William B. Karesh, *Appointment at the Ends of the World*, Warner Books, 1999

Kenn Kaufman, *Kingbird Highway*, Houghton Mifflin, 2006

Elizabeth Kolbert, *The Sixth Extinction: An Unnatural History*, Henry Holt, 2014

Mark Kurlansky, *Cod*, Walker and Co., 1997; *The Last Fish Tale*, Ballantine Books, 2008

Tim Laman and Edwin Scholes, *Birds of Paradise*, National Geographic Society, Cornell Lab of Ornithology, 2012

Richard Leakey and Roger Lewin, *The Sixth Extinction: Biodiversity and Its Survival,* Doubleday, 1995

Aldo Leopold, *A Sand County Almanac*, Oxford University Press, 1949

Richard Louv, *Last Child in the Woods*, Algonquin Books of Chapel Hill, 2005

Helen Macdonald, *H Is for Hawk*, Grove Press, 2014

Sallie McFague, *The Body of God*, Fortress Press, 1993

John McPhee, *Encounters with the Archdruid*, Farrar, Straus and Giroux, 1971

Farley Mowat, *Sea of Slaughter*, McClelland and Stewart, 1984

JoAnne M. Mowczko, *The Achuar of the Pastaza River*, self-published, 2014

Roger Tory Peterson, edited by Bill Thompson III, *All Things Reconsidered*, Houghton Mifflin, 2006

David Quammen, *The Song of the Dodo*, Scribner, 1996

Alan Rabinowitz, *Jaguar,* Arbor House, 1986; *Chasing the Dragon's Tail,* Doubleday, 1991; *Beyond the Last Village*, Island Press, 2001; *Life in the Valley of Death*, Island Press, 2008; *An Indomitable Beast*, Island Press, 2014

Marc Reisner, *Cadillac Desert*, Penguin, 1993

David Rothenberg, *Why Birds Sing*, Basic Books, 2005

Carl Safina, *Song for the Blue Ocean,* Henry Holt, 1998; *Beyond Words*, Henry Holt, 2015

George B. Schaller, *A Naturalist and Other Beasts,* Sierra Club Books, 2007; *Tibet Wild,* Island Press, 2012

Phoebe Snetsinger, *Birding on Borrowed Time*, American Birding Association, 2003

Rob Stewart with Evan Rosser, *Save the Humans*, Random House Canada, 2012

William Stolzenburg, *Heart of a Lion*, Bloomsbury, 2016

Alfred Russel Wallace, *A Narrative of Travels on the Amazon and Rio Negro,* 1853

Scott Weidensaul, *Living on the Wind*, North Point Press, 1999

Terry Tempest Williams, *Red*, Vintage, 2002

E. O. Wilson, *The Social Conquest of Earth,* Liveright, 2012

Steve Winter with Sharon Guynup, *Tigers Forever,* National Geographic Society, Panthera, 2013

Patricia Chapple Wright, *For the Love of Lemurs,* Lantern Books, 2014

Julie Zickefoose, *The Bluebird Effect,* Houghton Mifflin Harcourt, 2012

I also recommend all the books in Cemex corporation's nature and conservation series produced over the past twenty years. Authored by top scientists such as Conservation International's Russell A. Mittermeier and including thousands of photographs by leading conservation photographers, these books are both enlightening and beautiful.

Index

Page numbers in *italics* refer to illustrations.

Illustration Credits

All of the photographs in the book were taken by Jane Alexander, with her camera, with the exception of those listed below, which were given graciously by Steve Winter, Dubi Shapiro and Joan de la Malla, Luis Garrido, scientists Bruce Beehler and Alan Rabinowitz, and conservationist Caroline Gabel.

Page 5   Courtesy of Alan Rabinowitz

Page 40   Courtesy of Steve Winter

Page 52   Courtesy of Steve Winter

Page 115 Courtesy of Bruce Beehler

Page 121 Courtesy of Caroline Gabel

Page 160 Courtesy of Tom Lovejoy

Page 170 Courtesy of Bill Conway

INSERT ONE

Page 3   (bottom) Courtesy of Dubi Shapiro

Page 4   Courtesy of Bruce Beehler

Page 5   (bottom) © Luiz Garrido, courtesy of the Wildlife Conservation Society

Page 6   (bottom) Courtesy of Joan de la Malla

INSERT TWO

Page 3   (top) Courtesy of Joan de la Malla

Page 4   Courtesy of Bruce Beehler

Page 5   (top) Courtesy of Joan de la Malla

Page 6   (top) Courtesy of Joan de la Malla

Page 7   (bottom) Courtesy of David Yarnold

A Note on the Type

The text of this book was set in Simoncini Garamond, a modern version by Francesco Simoncini of the type attributed to the famous Parisian type cutter Claude Garamond (ca. 1480–1561). Garamond was a pupil of Geoffroy Tory and is believed to have based his letters on the Venetian models, although he introduced a number of important differences, and it is to him we owe the letter that we know as oldstyle. He gave to his letters a certain elegance and a feeling of movement that won for their creator an immediate reputation and the patronage of Francis I of France.

Composed by North Market Street Graphics,
Lancaster, Pennsylvania

Printed and bound by Berryville Graphics,
Berryville, Virginia

Designed by Betty Lew